UNIX AND SHELL PROGRAMMING

Devendra Kumar
Sr. Asst. Professor
ABES Engineering College, Ghaziabad

S.K. KATARIA & SONS®
Publisher of Engineering Books
4885/109, Prakash Mahal, Dr. Subhash Bhargav Lane,
Opposite Delhi Medical Association, Daryaganj, New Delhi–110002 (INDIA)
Phone: +91-11-43551243; Mobile: +91-9871775858, 7678425209
e-mail: katariabook@gmail.com, katariabooks@yahoo.com
Website: www.skkatariaandsons.com
(e-books, subscriptions and bulk buying also available on our website)

Published by:

S.K. Kataria & Sons®

4885/109, Prakash Mahal, Dr. Subhash Bhargav Lane,
Opposite Delhi Medical Association, Daryaganj, New Delhi-110002 (INDIA)
Phone: +91-11-43551243; Mobile: +91-9871775858, 7678425209
e-mail: katariabook@gmail.com, katariabooks@yahoo.com
Website: www.skkatariaandsons.com

First Edition: February 2008-09; Second Edition: January, 2011; Reprint: 2025

Laser Typesetting by : Plus Computers, Delhi

Preface.....

The inspiration, behind writing this book comes from my students who are very much impressed by my notes that are provided to them time to time. The course contents of UNIX and Shell Programming is scattered across many books and students found it difficult to purchase all the books. So I thought of putting the entire contents in single book for their convenience and guidance. This book covers the complete course contents of the subject UNIX and Shell Programming, which is the part of MCA 2nd Semester Curriculum, U.P. Technical University, Lucknow. We have tried our level best to add the latest literature available, and to make student aware of various UNIX and Shell Programming concept. This book is designed to help student in such a way that it starts with basic concept of UNIX Operating System and goes up to level to master the students. UNIX Operating System is the leading multi-user operating system in millions of users through out the world, and is gaining more and more popularity as the platform of choice together with linux.

The book is divided into 8 different chapters.

Chapter 1: **UNIX Introduction:** Operating System introduction, Type, functions, Unix overview, History of UNIX, UNIX system organization, kernel, shell and features of the UNIX

Chapter 2: **UNIX File System :** Definition of file and Directory, types of file, file system, file system organization: Boot Block, Super Block, I-node.

Chapter 3: **Editor :** Discuss about the editor (such as and ed).

Chapter 4: **General Purpose Utilities :** Discuss all type commands, internal external commands, and rules of commands.

Chapter 5: **UNIX Shell Programming :** Types of shell, wild cards, shell variables, shell commands etc. Shell script, control statements, looping, conditional statements trapping signal etc.

Chapter 6: **UNIX System Administration :** Introduction, mounting unmounting file system, system booting shutdown, handling user accounts, back up and recovery, system security.

Chapter 7: **UNIX Filter and Expression :** Filters using regular expression grep and sid; Advanced filters are part.

Chapter 8: **UNIX Tools and Debugger :** System Development Tools: Lint, make, and sccs, language development tools: Lex, yacc, and M4, Text Formatting Tools :troff, nroff,tbl, pic, and eqn, Debugger Tools: Dbx,Adb, Sdb, Strip, and Ctrace, UNIX system calls and Library functions.

I will be obliged to all those reader who will point out mistake and printing error to improve book. Suggestion for improvement will be most welcome through Email: **devendra_arya@rediffmail.com.** Finally, I wish all the reader of this book a very good luck and happy studying.

Devendra Kumar

✍ Acknowledgement.....

It is matter of great pleasure for me to present this second edition of the book **"UNIX and Shell Programming"** for the student of B.Tech (CS/IT) IVth sem and MCA Ist sem UKTU Dehradhun (UK) as well as MCA, BCA, B.SC (IT/CS) M.Sc (CS/IT) other universities. There are many people we would like to thanks for their help in writing this book. At the outset we thank our family and friends who inspired me to written this book. I am very much thankful to Sh. H.C. Mathur (Sr. Asso. Professor) in Department of Computer Applications ABES, Ghaziabad who gives me the guidelines and provided me good suggestion. I am also thanks to my colleagues who let us know which part of course material needs improvement. I will be thankful to Professor U.S Gupta former Head of the Department of Computer Applications, ABES, Ghaziabad and I am thankful to my A.B.E.S Engineering college Management, Prof. (Dr.) Surendra Kumar (Director). Prof. (Dr.) R.G Mendiratta (Director Academics), Professor Dinesh (Deputy Director), Prof. C.L. Mittal (Deam Academic), for valuable support.

I am very thankful to Professor D.Pandey (HoD C.S Department) ITS Engg. College, Gr. Noida, Professor Pradeep Sharma (Principal M.Tech) IEC Engg. College Gr. Noida, Prof. (Dr.) Rakesh Kumar (Director), K.S.J.I.M.T, Ghaziabad for genius help.

I am also thankful to all the faculty members of Department of Computer Applications ABES, Ghaziabad.

I would like to thank to my dear parent and my wife **Anita Arya** for her advice and encouragement during the completion of the book. I also thank to my younger brother **Harvendra Kumar** and my lovely Son Tanmay Arya for helpful comments and cooperation.

I would like to thank to my friends Er. Vineet Kumar (HCL Technology, Noida) Mr. Kuldeep Malik in ITS, Gr. Noida, Mr. Ishrat Ali in AIT, Ghaziabad, Mr.B.B. Das in KIET, Ghaziabad, Mr. Tarun Kr. Singhal in INMANTEC, Ghaziabad, Mr. Deepak Sexena in INMANTEC, Ghaziabad, Mr. Rama Krishna in IIMT, Gr. Noida, Mr. Prateek Mishra, HOD (MCA) SDCET, Ghaziabad Mr. Brahm Pal Singh at BIT Muzaffarnagar, Ms. Sunita Bisht in ITS Ghaziabad) and Mr. S. Goswami in IIMT Meerut Er. Amit Gupta Sopra Group, Noida, Ms. Upsana sharma, Amity University, Noida for their cooperation and support.

I would like to special thanks to **Mr. Sanjeev Kataria** for **M/s S. K. Kataria** & Sons Publisher and team member Mr. Sharda Singh, Mr. Manoj Tyagi, Mr. Vikas Rohilla, without the support of all can not success to written this book. Mr Kataria give me valuable suggestion and tips for written this book.

Devendra Kumar

✍ Contents.....

CHAPTER-1: AN INTRODUCTION TO UNIX 1-15

1.1 Introduction 1
1.2 Operating System 1
1.2.1 Types of Operating System 2
1.2.2 Functions of Operating System 3
1.3 Overview of UNIX 4
1.3.1 History of UNIX 6
1.3.2 Advantages and Disadvantages of UNIXx 9
1.4 UNIX System Organization 10
1.4.1 Kernel 10
1.4.2 Shell 11
1.4.3 Utilities 11
1.4.4 File System 11
1.4.5 Application Programs 12
1.4.6 Hardware 12
1.4.7 Users 12
1.5 Main Features of UNIX 12
1.6 Logging In and Out 13
1.6.1 Entering Your Userid and Password 13
1.6.2 Logging Out 14
1.6.3 Changing Your Password 14
■ Summary 15
■ Exercise 15

CHAPTER-2: UNIX FILE SYSTEM 16-28

2.1 Introduction 16
2.2 Files 16
2.2.1 Regular Files or Ordinary Files 17
2.2.2 Directory Files 17
2.2.3 Special Files or Device Files 17
2.3 Unix File Structure 17
2.3.1 Root Directory (/) 18
2.3.2 Basic File Attributes 18
2.4 File Systems 19
2.4.1 Types of File Systems 20
2.5 UNIX Directories 22
2.6 Unix File System Organization 23
2.6.1 Boot Block (BB) 23
2.6.2 Super Block (SB) 24
2.6.3 Data Block 24
2.6.4 Inode (Index Nodes) 24
2.7 Advantages and Disadvantages of UNIX File System Organisation 27
■ Summary 27
■ Exercise 28

CHAPTER-3: EDITORS (vi and ed) 29-76

3.1 Introduction 29
3.2 vi Modes 29
3.3 Getting Started 30
3.4 Moving the Cursor 32

3.5 Entering Text 33
3.6 Deleting Text 33
3.7 Replace and Substitute 34
3.8 Moving Text 35
3.9 Copying Text 35
3.10 Scrolling Text 36
3.11 Searching for Patterns in vi 37
3.12 Execute file, script or program 37
3.13 ed Editor 37
3.14 Flags 38
3.15 Pattern Matching 38
3.16 Regular Expressions 39
3.17 Forming Patterns 39
3.18 Restricting: Pattern matching 40
3.19 Addressing Lines 40
3.20 Subcommands 43
3.20.1 Editing a File 43
3.20.2 Manipulating Files 68
3.20.3 Miscellaneous Functions of the ed Editor Subcommands 70
3.21 Editing With ed 73
3.22 Features of ed 75
■ Summary 76
■ Exercise 76

CHAPTER-4: GENERAL PURPOSE UTILITIES 77-123

4.1 Introduction 77
4.2 Internal and External Commands 78
4.3 General Purpose Utilities 78
■ Summary 123
■ Exercise 123

CHAPTER-5: UNIX SHELL PROGRAMMING 124-168

5.1 Introduction 124
5.2 Types of Shells 125
5.2.1 Features of UNIX Shell 126
5.3 Shell Meta-Character (Wildcards) 127
5.3.1 The * Wildcard 127
5.3.2 The ? Wildcard 128
5.3.3 The Bracket ([ ]) Wildcard Character 128
5.3.4 Quoting Metacharacters 129
5.4 Shell Variables 129
5.4.1 Some Standard Shell Variables (Environment Variables) 131
5.4.2 Special Shell Variables 131
5.5 Shell Script 132
5.5.1 Read Statement 132
5.5.2 Simple Shell Script 133
5.6 Shell Commands 133
5.6.1 Positional Parameters: Set 133
5.6.2 Shifting Arguments: Shift 133
5.6.3 Changing Your Login Shell 134
5.7 The Environment 134
5.8 Integer Arithmetic 135
5.8.1 Arithmetic Expressions 136
5.9 Strings Manipulating 137
5.10 Special Command Line Characters 138

5.10.1 Conditional Execution Using && and || 138
5.10.2 Command Grouping: (Parentheses) 139
5.10.3 Inserting Comments 139
5.10.4 Double Quotes 139
5.10.5 Single Quotes 139
5.10.6 Back Quotes 139
5.10.7 Back Slash 140
5.10.8 Producing the Output 140
5.11 Advantage and Disadvantage of Shell Script 140
5.11.1 Advantages 140
5.11.2 Disadvantages 140
5.12 Decision making 141
5.12.1 The if-then- fi statement 141
5.12.2 The if-then-else-fi-statement 142
5.12.3 The if-then-elif-else-fi statement 146
5.12.4 The case- esac Statement 150
5.13 Looping Control 152
5.13.1 The for loop 152
5.13.2 The While Loop 153
5.13.3 The Until Loop 154
5.13.4 The Break Statement 155
5.13.5 The Continue Statement 155
5.14 Shell Operators 156
5.14.1 Arithmetic Operators 156
5.14.2 Relational Operators 157
5.14.3 Boolean Operators 157
5.14.4 String Operators 158
5.14.5 File Test Operators 158
5.15 Controlling Terminal Inputs 159
5.15.1 Output Redirection to a File 159
5.15.2 Input Redirection from a File 160
5.16 Trapping Signals 160
5.16.1 UNIX Signals 160
5.16.2 Handling Signals 161
5.16.3 Examples of Interrupt Handling 162
5.16.4 trap: Handling command lists 163
5.17 Array 164
5.17.1 Array Values 164
5.17.2 Accessing Array Values 164
5.18 Command Line Interpretation 166
5.19 Reserved Words 167
■ Summary 167
■ Exercise 168

CHAPTER-6: UNIX SYSTEM ADMINISTRATION 169-188

6.1 Introduction 169
6.1.1 Some Common System Administration Tasks 169
6.2 File System Mounting and Unmounting 170
6.2.1 The mount : Mount File System 171
6.2.2 The umount: Unmount File System 171
6.3 System Booting 172
6.3.1 System State Summary (or Init Level) 172
6.3.2 Boot Phases 172
6.4 Shutting down 173
6.5 Handling User Account 173
6.5.1 useradd: Adding new user 174

6.5.2 userdel & usermod: Removing and modifying users 175
6.5.3 Adding Group : groupadd 175
6.5.4 Deleting Group : groupdel 175
6.5.5 Modifying the Group: groupmod 176
6.6 Backup and Recovery 176
6.6.1 CPIO: Copy Input Output 176
6.6.2 The tar Command: the tape Archive Program 178
6.6.3 The dump Command 179
6.7 Security 180
6.7.1 Setting User ID (SUID) Permission 180
6.7.2 Passwd: Changing Password 181
6.7.3 The Sticky Bit 181
6.7.4 Chmod 181
6.7.5 Encrypting Files 182
6.8 Creating Files 182
6.9 Storage of Files 183
6.10 Disk Related Commands 184
6.10.1 df : For find the free Disk Space. 184
6.10.2 du : Used Disk Space 185
6.10.3 dfspace : Fee Disk Space int MB 185
6.10.4 ulimit : User Limit 186
6.11 User Quatas 186
6.12 User Accounts 187
■ Summary 188
■ Exercise 188

CHAPTER-7: UNIX FILTERS AND EXPRESSIONS 189-237

7.1 Introduction 189
7.2 Generalized Regular Expression Parser (GREP) 190
7.2.1 Searching a Pattern 190
7.2.2 Input for GREP 190
7.2.3 Options in GREP 191
7.3 The Stream Editor (SED) 192
7.3.1 Addresses 194
7.3.2 Substitution Command 194
7.3.3 Comments 194
7.3.4 Space Balls 195
7.4 AWK 195
7.4.1 awk Filtering 195
7.4.2 Splitting a Line into Fields 196
7.4.3 printf: Formatting Output 197
7.4.4 Variables and Expressions 198
7.4.5 The Comparison Operators 199
7.4.6 Variables 202
7.4.7 Arrays 203
7.4.8 Functions 204
7.4.9 Control Flow 206
7.5 PERL 212
7.5.1 Writing PERL Program 214
7.5.2 Data Types 215
7.5.3 Variables 218
7.5.4 Arrays 220
7.5.5 Operators 222
7.5.6 Functions 227
7.5.7 If Unless Statement 234
7.5.8 The for Statement 235

7.5.9 The while/until Statement 236
7.5.10 The foreach Statement 236
■ Summary 237
■ Exercise 237

CHAPTER-8: UNIX TOOLS 238-270

8.1 Introduction 238
8.2 Language Development Tools 238
8.2.1 Yacc-Yet another Compiler Compiler a Parser Generator 238
8.2.2 lex- Lexical Analyzer Generator 239
8.2.3 M4 242
8.3 Debugger Tools 242
8.3.1 Dbx 242
8.3.2 Adb (Absolute Debugger) 243
8.3.3 Sdb (Symbolic Debugger) 243
8.3.4 strip 243
8.3.5 Ctrace - C program Debugger 244
8.4 Library Functions 245
8.4.1 Finding Information about Library Functions 245
8.4.2 Use of Library Functions 245
8.4.3 Some Useful Library Functions 246
8.5 System Calls 246
8.5.1 Some Useful System Calls 246
8.6 System Development Tools 258
8.6.1 lint — C program checker (C- Program Verifier) 259
8.6.2 make 259
8.6.3 SCCS (Source Code Control System) 261
8.7 Text Formatting Tools 264
8.7.1 nroff (New run off) — format text 264
8.7.2 troff command 265
8.7.3 tbl Command 267
8.7.4 eqn Command 268
8.7.5 pic Command 269
■ Summary 270
■ Exercise 270

ADDITIONAL PROGRAMS (SCRIPTS) 271-283

MODEL QUESTION PAPERS 285-294

INDEX 295-299

CHAPTER 1

AN INTRODUCTION TO UNIX

1.1 INTRODUCTION

The UNIX operating system was designed to let a number of programmers access the computer at the same time and share its resources.

The operating system coordinates the use of the computer's resources, allowing one person, for example, to run a spell check program while another creates a document, lets another edit a document while another creates graphics, and lets another user format a document—all at the same time, with each user oblivious to the activities of the others.

The operating system controls all of the commands from all of the keyboards and all of the data being generated, and permits each user to believe he or she is the only person working on the computer.

This real-time sharing of resources make UNIX one of the most powerful operating systems ever. Although UNIX was developed by programmers for programming, it provides an environment so powerful and flexible that it is found in businesses, sciences, academia, and industry. Many telecommunications switches and transmission systems also are controlled by administration and maintenance systems based on UNIX.

While initially designed for medium-sized minicomputers, the operating system was soon moved to larger, more powerful mainframe computers. As personal computers grew in popularity, versions of UNIX found their way into these boxes, and a number of companies produce UNIX-based machines for the scientific and programming communities.

1.2 OPERATING SYSTEM

An operating system is a collection of program that acts as an interface between a user of a computer and computer hardware. A computer can store, process, and retrieve information; display multimedia document; search internet etc. through software.

A computer system can be roughly divided into four components:

(i) Hardware: Provide the basic computing resources.

(ii) The operating system: Interface between users and hardware.

(iii) The application programs: In which these resources are used to solve the problem of users.

(iv) The users: A user may be administrator, client or any other person who use the computer system.

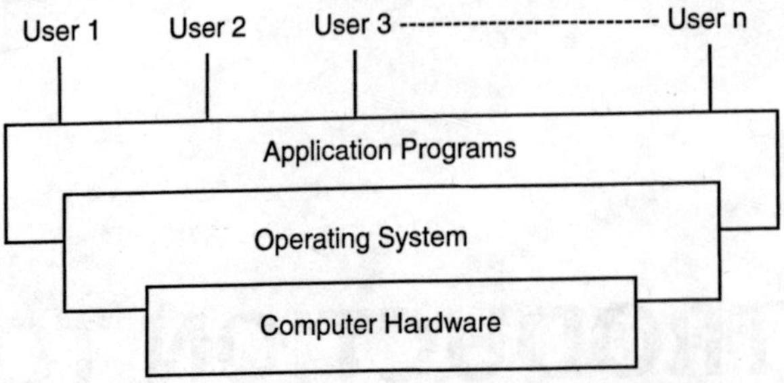

FIGURE 1.1 Computer System Component

The computer software can be divided roughly into two kinds:

(i) System programs: Which manage the operation of the computer itself.

(ii) Application programs: Which perform the actual work the user want.

The most fundamental computer system program is the operating system, which controls the computer's resources and provides the base upon which the application programs can be written.

1.2.1 Types of Operating System

Generally operating systems are of two types:

1. **Single-user operating system:** A single user operating system provides access to the computer system by a single user at a time. If another user needs access to the computer system, the user must wait till the current user finishes what they are doing and leaves. You might also have experienced this at home, where you want to use the computer but someone else is currently using it. You have to wait for them to finish before you can use the computer system.

   Operating system such as Windows 95, Windows NT workstation and Windows 2000 professional are essentially single user operating system.

2. **Multi-user operating system:** A multi-user operating system lets more than one user access the computer system at one time. Access to the computer system normally provided via network, so that users access the computer system remotely using a terminal or other computer. A multi-user operating system allows many different users to take advantage of the computer's resources simultaneously. In the early days of large multi-user computers, multiple terminals (keyboards and associated monitors) were provided. These terminals sent their commands to the main multi-user computer for processing, and the results were then displayed on the associated terminal monitor screen. Terminals were hard-wired directly to the multi-user computer system.

> ✍ **Note:** **Multi-Tasking Operating System:** A multi-tasking operating system provides the ability to run more than one program at once. For example, a user could be running a word processing package, printing a document, copying files to the floppy disk and backing up selected files to a tape unit. Each of these tasks the user is doing appears to be running at the same time.

A multi-tasking operating system has the advantage of letting the user run more than one task at once, so this leads to increased productivity. The disadvantage is that more programs that are run by the user, the more memory is required.

Operating System Interfaces

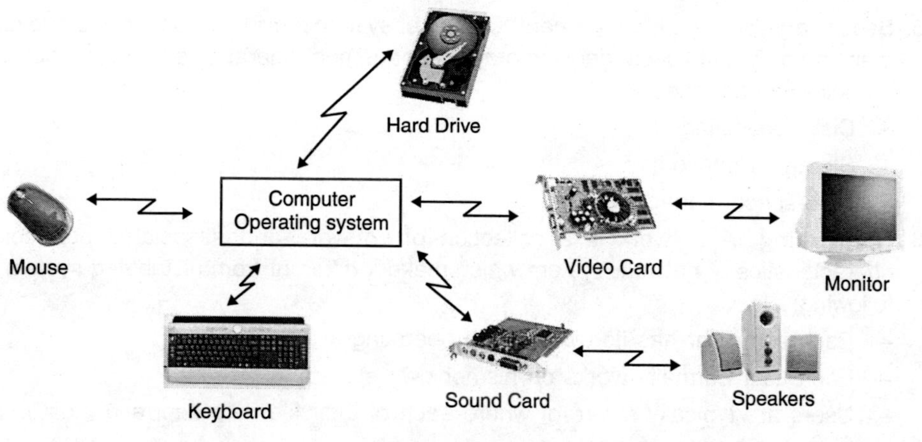

FIGURE 1.2

1.2.2 Functions of Operating System

The operating system usually provides many functions. These are given below:

1. **Process Management:** A process can be though of as a program in execution. A process needs some resources including CPU time, memory, files and Input/Output devices to complete its task. With process management operating system is responsible for the following activities:
   - The creation and deletion of user and system processes.
   - For process synchronization.
   - For process communication.
   - For Deadlock handling.
2. **Main-Memory Management:** Main memory is a large array of words or bytes.With main memory management operating system is responsible for the following activities:
   - Allocate and deallocate memory space as needs.
   - Decide which processes are to be loaded in to memory.
   - Keep record of the part of memory currently being used.

3. **File Management:** A file is a collection of related data or information defined by the creator. With file management operating system is responsible for the following activities:
   - Creation and deletion of files.
   - Backup of files.
   - Creation & deletion of directories.
   - Mapping files into secondary storage.
4. **Input/Output System:** The Input/Output consists of following:
   - Memory management including buffering, spooling and caching.
   - Device–driver interface.
   - Drivers for specific hardware devices.
5. **Secondary–Storage Management:** Computer system provide secondary storage to backup main memory. With secondary – storage management operating system is responsible for the following activities:
   - Disk scheduling.
   - Storage allocation.
   - Free-space management.
6. **Networking:** A network is a collection of software and associated protocol. Some characteristics of network system which make it different from distributed system are the following:
   - Each computer has its own private operating system.
   - Each user normally works on his/her own system.
   - Users are typically aware of where each of their files are kept and must move files from one system to another.
7. **Security:** Security refers to a mechanism for controlling the access of programs, processes or users to the resources defined by a computer system.
9. **Command–Interpreter system:-** Command interpreter is the interface between the user and the operating system. The command statements themselves deal with process creation and management, I/O handling, secondary–storage management, main memory management, file management, security and networking.

1.3 OVERVIEW OF UNIX

Unix is an operating system. An operating system tells a computer how to process commands and programs that the users give it. An operating system is the foundation software of a machine; it schedules tasks, allocates storage, and presents a default interface to the user between applications. Some other operating systems are DOS, Windows, BeOS, Linux, and the Macintosh OS.

Unix is a highly developed, grown-up, very stable, complex, and immensely powerful operating system. Unix provides multi-tasking, multi-user abilities that allow multiple programs to run on one computer simultaneously, and multiple users to use one computer simultaneously; while most operating systems only work on a specific type of machine. Unix has different flavors so that it can be run on many different types of machines. In addition to the operating system, there is a shell; the shell takes commands from the user and modifies them to suit the operating system. Since there are several different shells, this allows you to use Unix for different purposes, since each behaves in different ways. This is a powerful aspect of Unix, which is not shared by Windows or the Macintosh Operating Systems.

If you have used the DOS command line, some of what follows will be familiar. If you have mostly worked with graphical user interface systems such as Windows or Macintosh Operating Systems, it may take a while before you become comfortable working with Unix.

A computer needs special programming called software to make it work. UNIX systems have two kinds of software: the operating system software and the application or utility software. The operating system is the life of the computer.

Computer system is just a collection of metal, silicon, communication equipment and magnetic media. In the system some applications are stored which make the computer system useful to its users. UNIX is a large operating system. It has a large set of utility programs. These utilities include general user, programmer, type setting, communication and administrative commands.

It is a widely used multi-user operating system, originally developed by Bell Laboratories in 1969 to provide an environment in which along with the computer recourse management and services.

The hardware of the computer is made up of four basic parts, the Central Processing Unit (CPU), the memory, secondary storage (disks) and input/output (I/O) devices (terminals, printers, tapes).

These can do nothing without some software or instructions that can be executed by the CPU and tell it how to access other devices and process information received from them. The "operating system" is the basic software that is always accessible to the CPU (either always in memory or in a known location that can be loaded into memory as needed) that tells it how to access and control the resources of the computer.

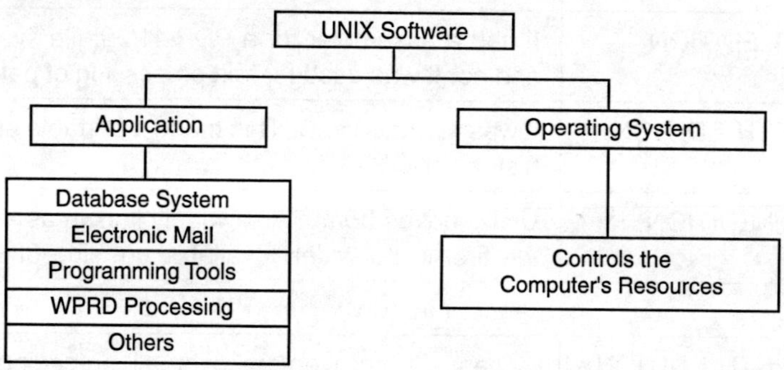

FIGURE 1.3 UNIX Overview

Main Operating System Activities:

- **Manages the physical resources of the computer**: Knows how to get data from the disk or talk to a terminal.
- **Arbitrates between processes and users:** Decides which process (program) gets the CPU or the use of peripherals, for how long, in what order.
- **Creates an environment to support user programs:** Provides a set of basic utilities so that programs can run without needing to know the detailed characteristics of the particular hardware. For example, the operating system provides routines to open a file and read it without having to know the specific geometry of the disk.

1.3.1 History of UNIX

In the late 1960s, researchers from General Electric, MIT and Bell Labs launched a joint project to develop an ambitious multi-user, multi-tasking OS for mainframe computers known as MULTICS (Multiplexed Information and Computing System). MULTICS failed, but it did inspire Ken Thompson, who was a researcher at Bell Labs, to have ago at writing a simpler operating system himself. He wrote a simpler version of MULTICS on a PDP7 in assembler and called his attempt UNICS (Uniplexed Information and Computing System). Because memory and CPU power were at a premium in those days, UNICS (eventually shortened to UNIX) used short commands to minimize the space needed to store them and the time needed to decode them—hence the tradition of short UNIX commands we use today, e.g. ls, cp, rm, mv etc.

Ken Thompson then teamed up with Dennis Ritchie, the author of the first C compiler in 1973. They rewrote the UNIX kernel in C—this was a big step forwards in terms of the system's portability and released the Fifth Edition of UNIX to universities in 1974. The Seventh Edition, released in 1978, marked a split in UNIX development into two main branches: SYSV (System 5) and BSD (Berkeley Software Distribution). BSD arose from the University of California at Berkeley where Ken Thompson spent a sabbatical year. Its development was continued by students at Berkeley and other research institutions. SYSV was developed by AT&T and other commercial companies.

Summarize the development of UNIX. This is given below:

1969	THE BEGINNING UNIX	The history of UNIX starts back in 1969, when Ken Thompson, Dennis Ritchie and others started working on the "little-used PDP-7 in a corner" at Bell Labs and what was to become UNIX.
1971	FIRST EDITION	It had a assembler for a PDP-11/20, file system, fork(), roff and ed. It was used for text processing of patent documents.
1973	FOURTH EDITION	It was rewritten in C. This made it portable and changed the history of OS's.
1975	SIXTH EDITION	UNIX leaves home. Also widely known as Version 6, this is the first to be widely available out side of Bell Labs.
1978	3BSD	Virtual Memory
1979	SEVENTH EDITION	It was a "improvement over all preceding and following Unices". Included the Bourne shell, UUCP, and C.
1980	Xenix	Microsoft introduces Xenix.
1980	4BSD	Introduced by UC Berkeley
1982	SYSTEM III	AT&T's UNIX System Group (USG) release System III, the first public release outside Bell Laboratories.
1983	SYSTEM V Release 1	AT&T announces UNIX System V, the first supported release.
1983	4.1BSD	UC Berkeley release with performance enhancements.
1984	4.2BSD	UC Berkeley releases with networking capabilities includes TCP/IP etc.

1984	System V Release 2	System V Release 2 introduced. For protection and locking files, and job control features added.
1986	HP-UX	First version of HP-UX released for HP Precision Architecture
1986	4.3BSD/ AIX Version	4.3BSD released, including internet name server. SVID introduced. NFS shipped. AIX announced.
1987	System V Release 3	System V Release 3 including STREAMS, TLI, RFS.
1988	POSIX	POSIX.1 published. Open Software Foundation (OSF) and UNIX International (UI) formed. Ultrix 4.2 ships.
1989	System V Release 4	UNIX System V Release 4 ships, unifying System V, BSD and Xenix.
1990	XPG3	X/Open launches XPG3 Brand. OSF/1 debuts.
1991	386BSD	Based on BSD for Intel 80386.
1991 1992	Linux 0.01 System V Release 4.2	Linux Torvalds started development of Linux. USL releases UNIX System V Release 4.2.
1992	HP-UX9.0	Supported workstations, inclding a GUI
1993	4.4BSD	4.4BSD the final release from Berkeley.
1993	FreeBSD1.0	Initial release based on 4.3BSD and 386BSD.
1993	System V Release 4.2	MP Novell transfers rights to the "UNIX" trademark and the Single UNIX Specification to X/Open.Last version of UNIX developed by USL.
1994	Single UNIX Specification	As the new owner of the UNIX trademark, X/Open introduces the Single UNIX Specification, separating the UNIX trademark from any actual code stream.
1994	Linux1.0	First Version of Linux.
1994	NetBSD1.0	First multiplatform release.
1994	Solaris	Multi Supported.
1995	UNIX 95	X/Open introduces the UNIX 95 branding program for implementations of the Single UNIX Specification.
1995	Solaris2.5	Common Desktop Environment Supported.
1995	HP-UX 10.0	Conformed to the single UNIX specification and CDE.
1996	Linux 2.0	Performance improvements.
1996	OpenBSD1.2	Initial release with strong support of security.
1997	Solaris2.6	UNIX 95 complaint set.

1997	Single UNIX Specification	Version 2 The Open Group introduces Version 2 of the Single UNIX Specification, including support for real time, threads and 64-bit and larger processors.
1997	System V Release 5	Increase reliability.
1997	Unix Ware7/HP-UX11.0	SCO UNIX based on SVR5 kernal.64-Bit OS.
1997	AIX 4.3	Support for 64-bit architectures, registered with UNIX 98.
1998	UNIX 98	The Open Group introduces the UNIX 98 family of brands, including Base, Workstation and Server. First UNIX 98 registered products shipped by Sun, IBM and NCR.
1998	FreeBSD3.0	Kernel changes and security fixes.
1998	Solaris 8	Performance and application support enhacements.
1999	Linux/Darwin/UNIX at 30	The UNIX system reaches its 30th anniversary. Linux 2.2 kernel released. The Open Group and the IEEE commence joint development of a revision to POSIX and the Single UNIX Specification.
2000	HP-UX 11i	Introduces operating environments.
2000	FreeBSD4.0	Networking and security enhancements.
2001	Single UNIX Specification Version 3/Linux 2.4/AIX 5L	Version 3 of the Single UNIX Specification unites IEEE POSIX, The Open Group and the industry efforts. Linux 2.4 kernel released. Introduced affinity for Linux.
2001	Mac OS X 10.1"Cheetah"/ "Puma"	Devices support, and software development environment./More complete than Cheetah, with performance enhancements and support.
2002	Solaris 9	Manageability, security and performance.
2003	ISO/IEC 9945:2003	The core volumes of Version 3 of the Single UNIX Specification are approved as an international standard. The "Westwood" test suite ship for the UNIX 03 brand. Solaris 9.0 E ships. Linux 2.6 kernel released.
2004	Solaris 10/ NetBSD 2.0	Advance performance, security and availability./ Support for SMP
2005	OpenServer 6	Improved SMP support.
2006	Open salaries (Build 52)	For open Group with advance features.
2007	Mac OS X 10.4.10 server	Apple Max OC X certified to UNIX 03.

2008	ISO/IEC 9945 : 2008	Latest revision of the UNIX API set formally standardized at ISO/IEC, IEEE and the open group. Adds further APIs.
2009	UNIX at 40	IDC on UNIX market-says UNIX $69 billion in 2008, Predicts UNIX $74 billion in 2013.
2010	AIX 7 open beta	Certified by open market.

1.3.2. Advantages and Disadvantages of UNIXx

1. Advantages

(a) Full multi-tasking with protected memory. Multiple users can run multiple programs each at the same time without interfering with each other or crashing the system.

(b) Very efficient virtual memory, so many programs can run with a modest amount of physical memory.

(c) Access controls and security. All users must be authenticated by a valid account and password to use the system at all. All files are owned by particular accounts. The owner can decide whether others have read or write access to his files.

(d) A rich set of small commands and utilities that do specific tasks well—not cluttered up with lots of special options. Unix is a well-stocked toolbox, not a giant do-it-all Swiss Army Knife.

(e) Ability to string commands and utilities together in unlimited ways to accomplish more complicated tasks—not limited to preconfigured combinations or menus, as in personal computer systems.

(f) A powerfully unified file system. Everything in a file: data, programs, and all physical devices. Entire file system appears as a single large tree of nested directories, regardless of how many different physical devices (disks) are included.

(g) A main kernel that does the basics for you but doesn't get in the way when you try to do the unusual.

(h) Available on a wide variety of machines—the most truly portable operating system.

(i) Optimized for program development, and thus for the unusual circumstances that are the rule in research.

2. Disadvantages

(a) The traditional command line shell interface is user hostile—designed for the programmer, not the casual user.

(b) Commands often have cryptic names and give very little response to tell the user what they are doing. Much use of special keyboard characters—little typos have unexpected results.

(c) To use Unix well, you need to understand some of the main design features. Its power comes from knowing how to make commands and programs interact with each other, not just from treating each as a fixed black box.

(d) Richness of utilities (over 400 standard ones) often overwhelms novices. Documentation is short on examples and tutorials to help you figure out how to use the many tools provided to accomplish various kinds of tasks.

1.4 UNIX SYSTEM ORGANIZATION

A UNIX structure is a computer operating system structure. UNIX operating system is made up of several major components. These components include the kernel, shell, utilities, system calls, users and application programs. The relationship among the users, shell, kernel, application programs and hardware are given below.

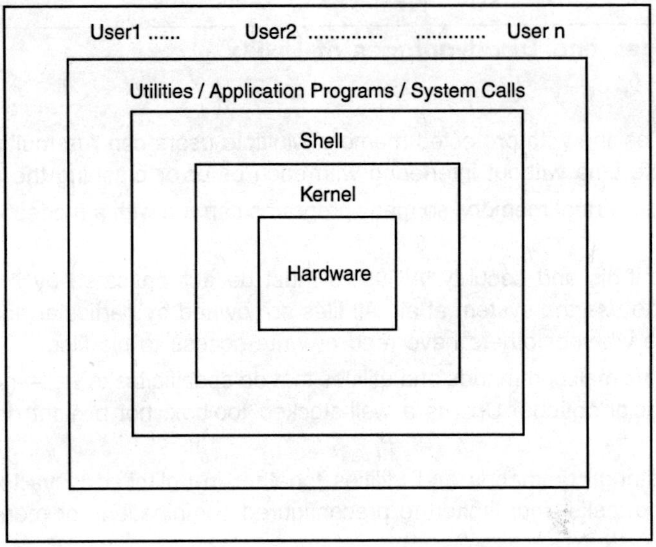

FIGURE 1.4 UNIX system architecture

1.4.1. Kernel

A Unix kernel is the core or key of the operating system. The kernel is the central component of most computer operating systems. Its responsibilities include managing the system's resources the communication between hardware and software components as the basic component of an operating system. The kernel provides the low-level abstraction layer for the resources (especially memory, processors and I/O devices) that application software must control to perform its function. The kernel makes these facilities available to application processes through inter-process communication mechanisms and system calls. The kernel consists of many kernel subsystems like process management, memory management, file management, device management and network management.

Each of the subsystem has some features:

(i) **Concurrency:** As Unix is a multi-processing OS, many processes run concurrently to improve the performance of the system.

(ii) **Virtual Memory (VM):** Memory management subsystem implements the virtual memory concept and a user need not worry about the executable program size and the RAM size.

(iii) **Paging:** It is a technique to minimize the internal as well as the external fragmentation in the physical memory.

(iv) **Virtual File System (VFS):** A VFS is a file system used to help the user to hide the different file systems complexities. A user can use the same standard file system related calls to access different file systems.

The kernel provides these and other basic services: interrupt and trap handling, separation between user and system space, system calls, scheduling, timer and clock handling, file descriptor management.

Features of Kernel

Some key features of the UNIX architecture concept are:

1. UNIX systems utilize a centralized operating system kernel which manages system and process activities.
2. All non-Kernel software is organized into separate, kernel-managed processes.
3. UNIX systems are preemptively multi-tasking: multiple processes can run at the same time, or within small time slices and nearly at the same time, and any process can be interrupted and moved out of execution by the kernel. This is known as thread management.
4. Files are stored on disk in a hierarchical file system, with a single top location throughout the system (root, or "/"), with both files and directories, subdirectories, sub-subdirectories, and so on below it.
5. With few exceptions, all devices and some types of communications between processes are managed and visible as files or pseudo-files within the file system hierarchy. This is known as everything's a file.

1.4.2. Shell

The shell read your commands and interprets them as request to execute a program or programs, which it then arranges to have carried out. Because once you have logged in, you are talking to a program called the shell. Shell is the command line interpreter. Users interact with system through shell. The shell interprets the commands that you type, generally acting as the user interface to the UNIX operating system. Many different UNIX shells are available. Shell is command interpreter. It is actually the interface between the user and kernel.

The three most popular shells are:

1. **Bourne Shell:** The Bourne shell, the most compact shell but also the simplest.
2. **C Shell:** The C shell, which uses C syntax and has many conveniences.
3. **Korn Shell:** The Korn shell, a superset of the Bourne shell that lets you edit the command line.

1.4.3. Utilities

Utilities are executable programs for the users to perform a variety of standard functions supplied with the system. The UNIX system contains several hundred utilities or user programs. Commands are also called tools, because they can be used separately or put together in various ways to carry out useful tasks. You execute these utilities by invoking them by name through the shell. That's why they are called commands.

1.4.4. File System

To use a Unix system, the user has to perform login process. UNIX operating system is the concept of directories and files. In UNIX, the directories have names and are arranged in a hierarchical sequence with the top directory being the "root" directory, known as "/" ("/" directory, there are subdirectories (e.g. **user** and **home**). The UNIX file system provides a logical method for organizing, storing, retrieving, managing and manipulating information. In any directory, there can also be "files". Files can be "readable", "writable" and/or "executable". Files are organized into a hierarchical file system, with files grouped together into directories.

1.4.5. Application Programs

Applications are user created software and third party purchased software. We can use applications built using UNIX commands, tools and programs. Application programs carry out many different types of tasks. Some tasks perform general functions that can be used by a variety of users in industry, government and education. These are known as horizontal applications and include such programs as compilers, DBMS (Database Management System), word processors and communication programs. In fever of industry – specific are known as vertical applications. For examples include software packages used for managing a running bank, a hotel and operating point of sale terminals.

1.4.6 Hardware

The hardware provides the basic computing resources. The resources are central processing unit, the memory, and the input/output (I/O) devices.

1.4.7 Users

The operating system controls and coordinates the use of hardware among the various applications programs for various users.

1.5 MAIN FEATURES OF UNIX

1. **Multi-user:** More than one user can use the machine at a time supported via terminals (serial or network connection). A multi-user operating system allows more than one user to share the same computer system at the same time. It does this by time-slicing the computer processor at regular intervals between the various users. This switching between user programs is done by part of the kernel. To switch from one program to another requires
   - a regular timed interrupt event (provided by a clock)
   - saving the interrupted programs state and data
   - restoring the next programs state and data
   - running that program till the next timed interrupt occurs
2. **Multi-tasking:** More than one program can be run at a time Multi-tasking operating systems permit the use of more than one program to run at once. It does this in the same way as a multi-user system, by rapidly switching the processor between the various programs. Multi-tasking systems support foreground and background tasks. A foreground task is one that the user interacts directly with using the keyboard and screen. A background task is one that runs in the background (it does not have access to the keyboard). Background tasks are usually used for printing or backups.
3. **Hierarchical directory structure:** To support the organization and maintenance of files. Supports hierarchical file system to hold user data organized in the form of directories and files.
4. **Portability:** A major contribution of the UNIX system was its portability, permitting it to move from one brand of computer to another with a minimum of code changes. At a time when different computer lines of the same vendor didn't talk to each other—yet alone machines of multiple vendors—that meant a great savings in both hardware and software upgrades.Only the kernel (<10%) written in assembler tools for program development a wide range of support tools (debuggers, compilers)

5. **Multi-programming:** Support more than one program, in memory, at a time. Amounts to multiple user processes on the system.
6. **Communications:** E-mail is common today, but it has only come into its own in the business community within the last 10 years. Not so with UNIX users, who have been enjoying e-mail for several decades.

   UNIX e-mail at first permitted users on the same computer to communicate with each other via their terminals. Then users on different machines.UNIX systems around the world were linked into a world wide web decades before the development of today's World Wide Web.
7. **UNIX Tools:** UNIX comes with hundreds of programs that can divided into two classes:
   - **Integral utilities** that are absolutely necessary for the operation of the computer, such as the command interpreter, and
   - **Tools** that aren't necessary for the operation of UNIX but provide the user with additional capabilities, such as typesetting capabilities and e-mail.
8. **Documentation:** There are several newsgroups on UNIX where you can put your queries in case you are stranded with problem. In UNIX we have a shell programming or a network configuration issue. The frequently asked questions are a document that addresses common problems widely available on the net. UNIX is easily tamed today.
9. **Security:** Identifies a user with a userid and groupid and allows access permissions to resources to be specified using these ids.
10. **Command Language:** Supports a command language selectable on a per user basis. (Example: csh, sh and ksh)
11. **Utilities:** Supports a large number of tools, libraries and utilities to aid software development.

1.6 LOGGING IN AND OUT

Getting the Login Prompt

Before you can start using the system you must login to it. The method that you use to login varies depending on the type of device that you are using to login.

1.6.1 Entering Your Userid and Password

In order to login, you MUST have an account on the system you are accessing. Remember, different domains require different accounts. When you picked up your account information, you were given a userid and a password. This combination of information allows you to access your account. Type your userid using lower-case letters, then press the ENTR key. It is very important that you use lower-case letters for your userid. If you make a typing mistake, you can correct it by pressing the DELETE key once for each character you wish to erase. You must make your corrections before you press the ENTR key.

After you have entered your userid, the system will prompt you for your password (by displaying the word "Password:" if it is not already on the screen, or by moving the cursor behind the word "Password:" already on your screen). Enter your password and press the ENTRY key. Notice that the system does not show or "echo" your password as you type it. This prevents other people from learning your password by looking at your screen.

If you receive a message similar to "Login failed, please try again," you may have typed your userid or password incorrectly. Try again, making sure to type in your userid and password correctly

Login:

Password:

For example

```
Login:kumar
Password:****
$
```

The system will then display the command prompt. The prompt signals that the system is ready for you to enter your next command. Once you finish typing a command, you must always press ENTER to execute it.

System display the command prompt

$

1.6.2 Logging Out

To end a work session, you must explicitly log out of a UNIX session. To do this, type logout at the command prompt. Once you have logged out, the system will either display the login prompt again or begin executing a screen saver program.

To log out of the UNIX session type the command exit in front of command prompt, and press ENTER. If you try to use the **logout** command in the UNIX, you will receive the message, "Not in login shell."

For example

```
$ exit [Enter]
  Or
$ ctl + d
```

1.6.3 Changing Your Password

You can change your password at any time. You should change it the first time that you log in, and we recommend that you change it on a regular basis. At the command prompt, type **passwd**.You will be prompted to enter your old password and be asked twice to enter your new password. Neither your old nor new password will appear on the screen as you type. In order to be accepted, your password must meet the following conditions:

1. It must be at least seven characters long.
2. It must not match anything in your UNIX account information, such as your login name, or an item from your account information data entry.
3. It must not be found in the system's spelling dictionary unless a character other than the first is capitalized. It must not have three or more consecutively repeated characters or words in the dictionary contained within it.

For example, changing your password from arya80 to dk80 will look similar to the following example, except that the keystrokes for you old and new password will not be echoed on the screen.

```
password
   current password: arya80
   New password (? for help): dk80
```

```
New password (again): dk80
Password changed for userid
```

☞ SUMMARY

We have learned about the structure and components of UNIX. We will find this background information useful as we move to chapters 2, 3, and 5, where we will learn how to the basic features and capabilities of UNIX, such as files and directories, basic commands and the shell. This chapter has described the birth, history and evolution of UNIX. We learned about the evoluation of UNIX system V, the BSD (Berkeley Software Distribution), Linux and GNU. Then we become acquainted with the modern history of UNIX, including descriptions of the important standards in the UNIX world. This chapter then covered the origins and many important UNIX variants, including Linux, BSD, AIX, HP-UX and others. This chapter briefly explored the possible future of UNIX and proffered advice on selecting a UNIX variant that can meet our needs. We have also learn briefly about the operating system as well as operating system functions.

☞ EXERCISE

1. Draw the architecture of UNIX system and describe each layer. (UPTU 2001, 02, 04, 05)
2. What are the services of kernel in time shared manner? (UPTU 2001)
3. What do you mean by KERNEL in Unix Operating System? Describe in brief the main features of KERNEL. (UPTU 2002, 04)
4. Differentiate between multi-tasking and multi-user system. (UPTU 2002)
5. What are the important features of UNIX operating system? (UPTU 2004)
6. What is Kernel? What function perform by kernel.
7. "UNIX views everything in the form of file". Justify this statement with suitable examples.
8. Why is UNIX more portable than any other operating system? Discuss its multi-user and multi-tasking features.
9. What is a 'UNIX like' operating system?
10. What is in version 3 of the single UNIX specification?

❑❑❑

CHAPTER 2

UNIX FILE SYSTEM

2.1 INTRODUCTION

The UNIX file system provides a flexible and powerful way to manage and organize your information. In this chapter we will discus the basic concept of file and directory.

2.2 FILES

A file is a named collection of related data that appears to the user as a single, contiguous block of information and that is retained in *storage*.

Storage refers to computer devices or media which can retain data for relatively long periods of time, such as Hard Disk Drives (HDDs), CDROMs and magnetic tape. This contrasts with *memory*, which consists of RAM (Random Access Memory) chips and whose contents can be *accessed* (i.e., read and written to) at extremely high speeds but which are retained only temporarily. Every file has a title called filename. A filename can have almost any sequence of characters, up to 255 characters. Two filenames are considered the same if the first 14 characters are identical.

Types of Files

There are three basic types of files:

(i) regular : Stores data (text, binary, and executable).

(ii) directory : Contains information used to access other files.

(iii) special : Defines a FIFO (first-in first-out) pipe file or a physical device.

All file types recognized by the system call into one of these categories. However, the operating system uses many variations of these basic types.

2.2.1 Regular Files or Ordinary Files

Regular files are the most common files. Another name for regular files is ordinary files. Regular files contain data.

Regular file further divided into two categories.

1. Text Files

Text files are regular files that contain information readable by the user. This information is stored in ASCII. You can display and print these files. The lines of a text file must not contain **NULL** characters, and none can exceed [LINE_MAX] bytes in length, including the new-line character.

The term *text file* does not prevent the control of inclusion other non-printable characters (other than **NULL**). Therefore, standard utilities that list text files as inputs or outputs are either able to process the special characters gracefully or they explicitly describe their limitations within their individual sections.

2. Binary Files

Binary files are regular files that contain information readable by the computer. Binary files may be executable files that instruct the system to accomplish a job. Commands and programs are stored in executable, binary files. Special compiling programs translate ASCII text into binary code.

The only difference between text and binary files is that text files have lines of less than **[LINE_MAX]** bytes, with no **NULL** characters, each terminated by a new-line character.

2.2.2 Directory Files

Directory files contain information the system needs to access all types of files, but they do not contain the actual file data. As a result, directories occupy less space than a regular file and give the file system structure flexibility and depth. Each directory entry represents either a file or a subdirectory. Each entry contains the name of the file and the file's index node reference number (i-node). The i-node points to the unique index node assigned to the file. The i-node describes the location of the data associated with the file. Directories are created and controlled by a separate set of commands.

2.2.3 Special Files or Device Files

Special files define devices for the system or temporary files created by processes. There are three basic types of special files:

1. **FIFO** (first-in first-out), block, and character. FIFO files are also called pipes.
2. **Pipes** are created by one process to temporarily allow communication with another process. These files cease to exist when the first process finishes.
3. Block and character **files** define **devices.**

Every file has a set of permissions (called access modes) that determine who can read, modify, or execute the file.

2.3 UNIX FILE STRUCTURE

UNIX has a tree-like file system starting at the highest level with a directory called root (referred to as /). A directory on Unix is similar to a folder on a microcomputer. The structure and content of many of the directories is similar on various UNIX implementations. That is, the user does not decide how to set up all of the directories. There is a preset meaning to certain directories. Directories like bin, usr and etc. contain certain system level information and are maintained by a system administrator. The figure below displays the uppermost levels of some of the directories.

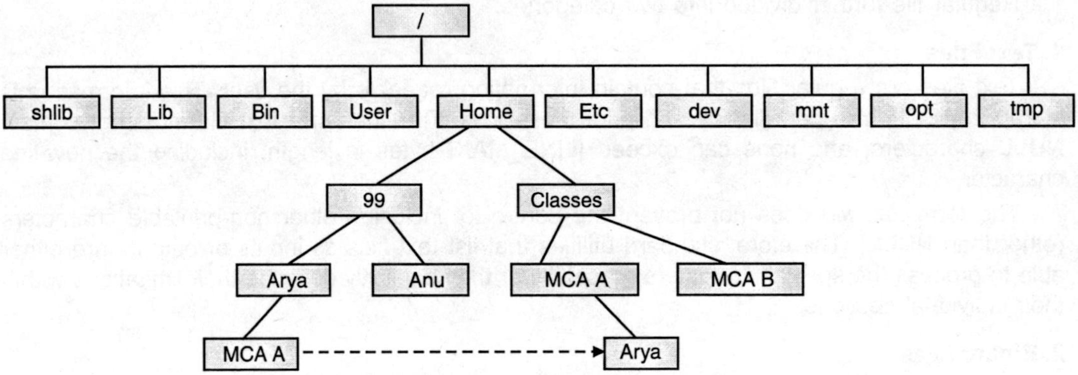

FIGURE 2.1 UNIX File Structure

The directory you will be most interested in is your home directory and that will be found somewhere under the /home directory path.

2.3.1. Root Directory (/)

The root directory is the directory on Unix operating systems that contains all other directories and files on the system and which is designated by a forward slash (/).

The use of the word root in this context derives from the fact that this directory is at the very top of the directory tree diagram that is commonly used to represent a filesystem. A filesystem is the hierarchy of directories that is used to organize directories and files on a computer.

The rooty directory contains the following system directories :

`/bin`	UNIX commands
`/dev`	Device special directory
`/etc`	Additional programs and data files
`/lib`	C programmiong library directory
`/mnt`	Mount directory; usually reserved for mounting file systems
`/opt`	Location of shared Software Storage Objects (SSO's)
`/shlib`	Shared libraries
`/stand`	Contains the kernel (unix) and boot files, including the loader (boot)
`/tcb`	The trusted computing base security binaries and databases
`/tmp`	Temporary directory
`/usr`	User routines
`/var`	Non-shared software storage object files

2.3.2. Basic File Attributes

UNIX files also have a set of file permissions associated with them. These permissions and file information's are called file attributes. This makes it possible for you to share files with others. Everyone on the same computer or only to a certain group of people and you can specify the level of access you want to grant (read, write, and execute). Read access allows you to view files but not change them. Write access allows you to change the contents of the file. Execute permission lets you run (execute) the file.

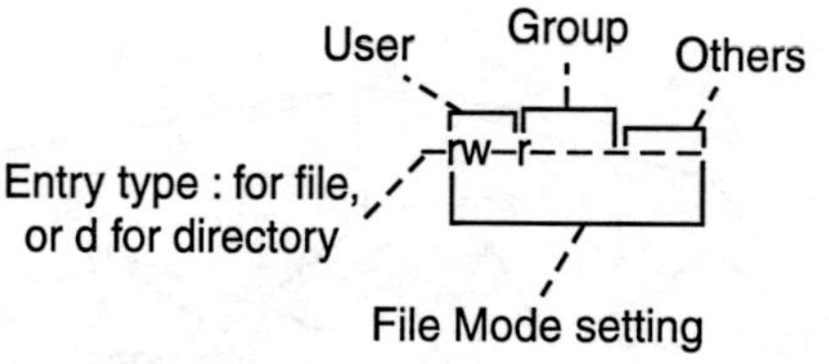

FIGURE 2.2 Diagram of UNIX File Architecture

After the first character, this identifies the entry type i.e. d for a directory and - for a regular file. You will find exactly nine (9) other characters. In order to make sense of these, you have to break them up in your head, into three groups of three symbols each (3 times 3 = 9).

The first three symbols represent the access permissions/attributes which apply to the user who owns the file i.e. owner.

The second group of three symbols represents the access permissions/attributes which apply to the group associated with the file. In UNIX, every userid is a member of one or more groups, and separate permissions/attributes can be set for the group, as opposed to the owner.

The third and last group of three symbols represents the access permissions and attributes which apply to all other userids other than the owning user and members of the owning group.

Now, let's look at the individual sets of attributes. Each of the three sets such as User, Group, and Others is identical, with respect to the meaning of the three characters of which it is composed.

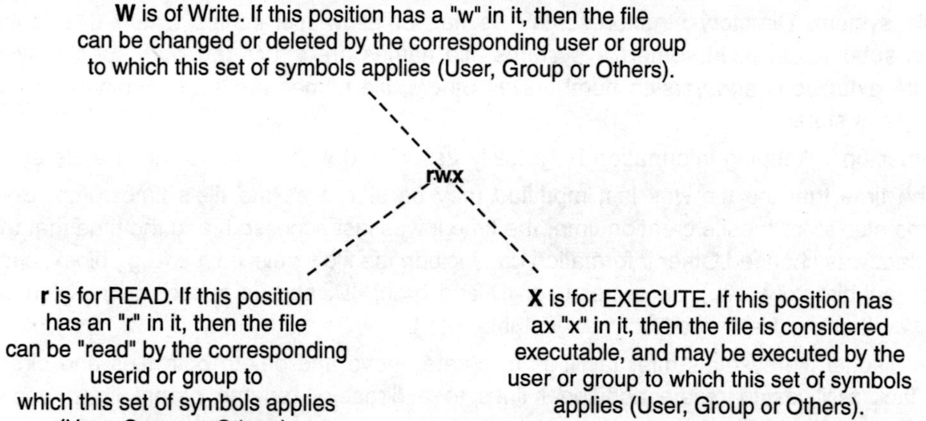

FIGURE 2.3 UNIX File-Access Attributes

2.4 FILE SYSTEMS

The most familiar file systems make use of data storage device that offers access to an array of fixed-size blocks, sometimes called *sectors*, generally 512 bytes each. The file system software is responsible for organizing these sectors into files and directories, and keeping track of which sectors belong to which file and which are not being used.

File systems need not make use of a storage device at all. A file system can be used to organize and represent access to any data, whether it be stored or dynamically generated. For example, from a network connection.

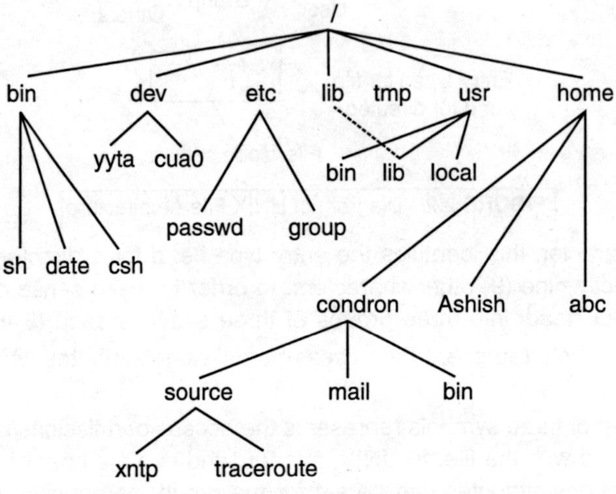

FIGURE 2.4 UNIX File System

Whether the file system has a storage device or not, file systems typically have directories which associate file names with files, usually by connecting the file name to an index into a file allocation table of some sort, such as the FAT in an MS-DOS file system, or an Inode in a Unix-like file system. Directory structures may be flat, or allow hierarchies where directories may contain subdirectories. In some file systems, file names are structured, with special syntax for filename extensions and version numbers. In others, file names are simple strings, and per-file metadata is stored.

Other book-keeping information is typically associated with each file within a file system.

The time that the file was last modified may be stored as the file's timestamp. Some file systems also store the file creation time, the time it was last accessed, and the time that the file's meta data was changed.Other information can include the file's device type (e.g., block, character, socket, subdirectory, etc.), its owner user-ID and group-ID, and its access permission settings (e.g., whether the file is read-only, executable, etc.).

Traditional file systems offer facilities to create, move and delete both files and directories. They lack facilities to create additional links to a directory rename parent links, and create bidirectional links to files.

Arbitrary attributes can be associated on advanced file systems, such as XFS, ext2/ext3, some versions of UFS, and HFS+, using extended file attributes. This feature is implemented in the kernels of Linux.

2.4.1. Types of file systems

File system types can be classified into disk file systems, network file systems and special purpose file systems.

1. Disk File Systems

A *disk file system* is a file system designed for the storage of files on a data storage device, most commonly a disk drive, which might be directly or indirectly connected to the computer.

Examples of disk file systems include FAT, FAT32, NTFS, HFS and HFS+, ext2, ext3, ISO 9660, ODS-5, and UDF.

2. Flash File Systems

A *flash file system* is a file system designed for storing files on flash memory devices. These are becoming more prevalent as the number of mobile devices is increasing, and the capacity of flash memories catches up with hard drives.

While a block device layer can run emulate hard drive behaviour and store regular file systems on a flash device, this is suboptimal for several reasons:

(i) **Erasing blocks:** Flash memory blocks have to be explicitly erased before they can be written to. The time taken to erase blocks can be significant, thus it is beneficial to erase unused blocks while the device is idle.

(ii) **Random access**: Disk file systems are optimized to avoid disk seeks whenever possible, due to the high cost of seeking. Flash memory devices impose no seek latency.

(iii) **Wear levelling:** Flash memory devices tend to "wear out" when a single block is repeatedly overwritten; flash file systems try to spread out writes as evenly as possible.

Such file systems include JFFS2 and YAFFS.

3. Database File Systems

A new concept for file management is the concept of a database-based file system. Instead of, in addition to, hierarchical structured management, files are identified by their characteristics, like type of file, topic, author, or similar metadata.

Example: **dbfs.**

4. Transactional File Systems

Each disk operation may involve changes to a number of different files and disk structures. In many cases, these changes are related, meaning that it is important that they all be executed at the same time. Take for example a bank sending another bank some money electronically. The bank's computer will "send" the transfer instruction to the other bank and also update its own records to indicate the transfer has occurred. If for some reason the computer crashes before it has had a chance to update its own records, then on reset, there will be no record of the transfer but the bank will be missing some money.

Transaction processing introduces the guarantee that at any point while it is running, a transaction can either be finished completely or reverted completely. This means that if there is a crash or power failure, after recovery, the stored state will be consistent.

This type of file system is designed to be fault tolerant, but may incur additional overhead to do so.

5. Network File Systems

A network file system is a file system that acts as a client for a remote file access protocol, providing access to files on a server.

Examples of network file systems include clients for the NFS, SMB protocols, and file-system-like clients for FTP and WebDAV.

6. Special Purpose File Systems

A special purpose file system is basically any file system that is not a disk file system or network file system. This includes systems where the files are arranged dynamically by software, intended for such purposes as communication between computer processes or temporary file space.

Special purpose file systems are most commonly used by file-centric operating systems such as Unix.

Examples include the procfs (/proc) file system used by some Unix variants, which grants access to information about processes and other operating system features.

Creating File

The cat is one of the most important command of UNIX. By using this command we can create file display content of the file and concatenate of more than two files.

```
$ cat >file1
This is Arya.
Press ctl + d
$
$cat >file2
From Ghaziabad.
Pres ctl + d
$
```

Display content

```
$ cat file1
This is Arya.
$
```

Concatenate of files

```
$cat file1 file2 > file3
$
$cat file3
This is Arya.
From Ghaziabad.
$
```

2.5 UNIX DIRECTORIES

UNIX directories are similar to regular files; they both have names and both contain information. Directories, however, contain other files and directories.

Many of the same rules and commands that apply to files also apply to directories.

All files and directories in the UNIX system are stored in a hierarchical tree structure. Envision it as an upside-down tree, as in the figure below.

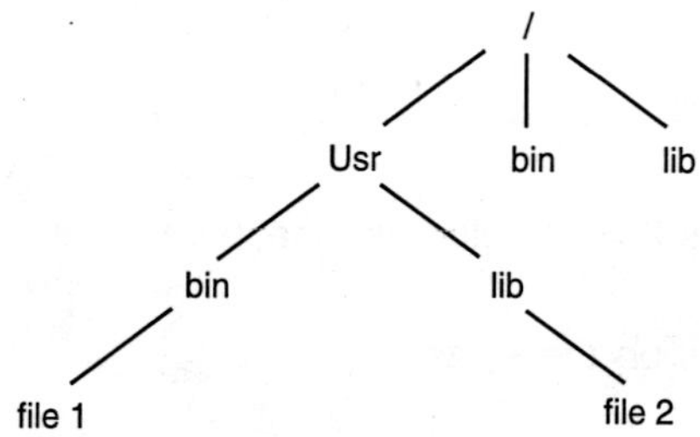

FIGURE 2.5 UNIX Directory Structure

At the top of the tree is the root directory. Its directory name is simply / (a slash character). Below the root directory is a set of major subdirectories that usually include **bin**, **dev**, **etc**, **lib**, **pub**, **tmp**, and **usr**. For example, the **/bin** directory is a subdirectory, or "child," of / (the root directory). The root directory, in this case, is also the parent directory of the **bin** directory. Each path leading down, away from the root, ends in a file or directory. Other paths can branch out from directories, but not from files.

Many directories on a UNIX system have traditional names and traditional contents. For example, directories named **bin** contain binary files, which are the executable command and application files. A **lib** directory contains library files, which are often collections of routines that can be included in programs by a compiler. **dev** contains device files, which are the software components of terminals, printers, disks, etc. **tmp** directories are for temporary storage, such as when a program creates a file for something and then deletes it when it is done. The **etc** directory is used for miscellaneous administrative files and commands. **pub** is for public files that anyone can use, and **usr** has traditionally been reserved for user directories, but on large systems it usually contains other **bin**, **tmp**, and **lib** directories.

Your *home directory* is the directory that you start out from when you first login. It is the top level directory of your account. Your home directory name is almost always the same as your userid.

2.6 UNIX FILE SYSTEM ORGANIZATION

In the original Unix file system, Unix divided physical disks into logical disks called *partitions*. Each partition is a stand alone file system. We will use the term "file system" when referring to a single partition.

Each disk device is given its own *major device number*, and each partition has an associated *minor device number* which the device driver uses to access the raw file system.

The major/minor device number combination serves as a handle into the device switch table. That is, the major number acts as an index, and the minor number is passed as an argument to the driver routines so that they can recognize the specific instance of a device.

Each file system contains:

2.6.1 Boot Block (BB)

A *boot block* located in the first few sectors of a file system. The boot block contains the initial bootstrap program used to load the operating system.

The first sector contains a bootstrap program that reads in a larger bootstrap program from the next few sectors, and so forth.

2.6.2 Super Block (SB)

A *super block* describes the state of the file system: the total size of the partition, the block size, pointers to a list of free blocks, the inode number of the root directory, magic number, etc.

The **super block** contains the following information, to keep track of the entire file system.

(i) Size of the file system
(ii) Number of free blocks on the system
(iii) A list of free blocks
(iv) Index to next free block on the list
(v) Size of the inode list
(vi) Number of free the inodes
(vii) A list of free inodes
(viii) Index to next free inode on the list
(ix) Lock fields for free block and free inode lists
(x) Flag to indicate modification of super block

2.6.3 Data Block

The **data blocks** contain the actual data contained in the files or directories. In Unix, a directory is a special file. A directory file contains names of the subdirectories and files present in that directory and its corresponding inode number.

Data blocks blocks containing the actual contents of files

Representation of Boot Block, Super Block, Inode and Data block in the file system

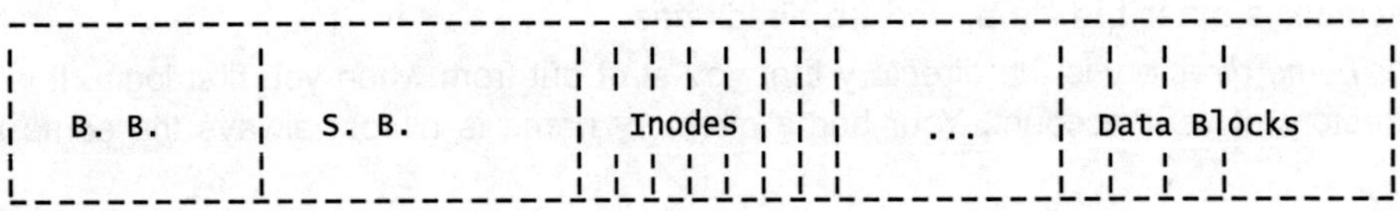

2.6.4 Inode (Index Nodes)

A linear array of *inodes* (index nodes). There is a one to one mapping of files to inodes and vice versa. An inode is identified by its "inode number", which contains the information needed to find the inode itself on the disk

Thus, while users think of files in terms of file names, Unix thinks of files in terms of inodes.

The **I-node list** (which server the purpose as FAT+ directory entries in DOS) is a list of inodes, which contains the following entries.

(i) Owner
(ii) Type
(iii) Last modified time
(iv) Last accessed time
(v) Last inode modified time
(vi) Access Permissions
(vii) No of links to the file
(viii) Size of the file
(ix) Data blocks owned

Owner indicates who owns the file(s) corresponding to this inode.

Type indicates whether inode represents a file, a directory, a FIFO, a character device or a block device. If the type value is 0 then the inode is free.

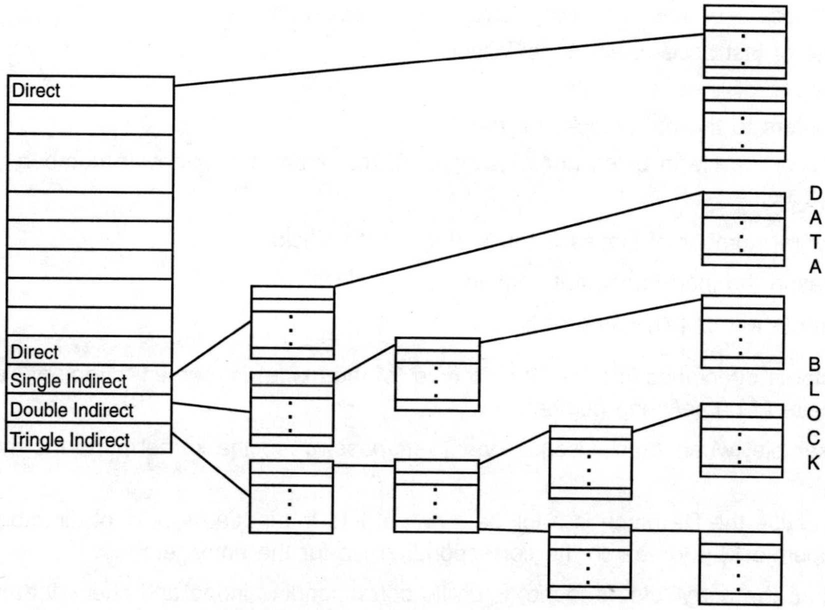

FIGURE 2.6 Inode DataBlock representation

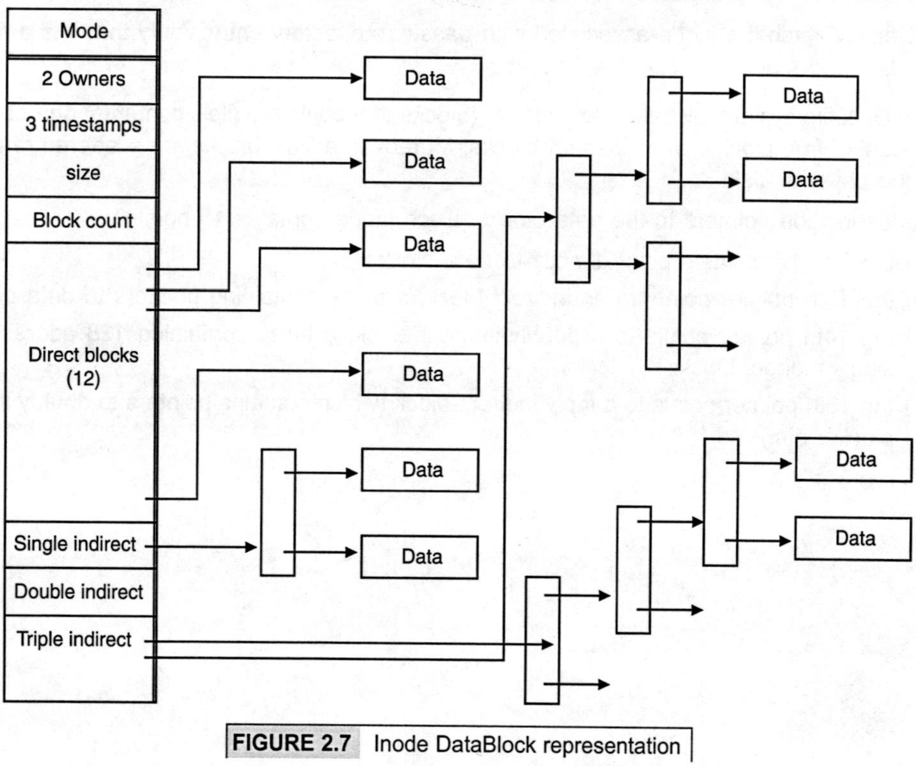

FIGURE 2.7 Inode DataBlock representation

An inode is the "handle" to a file and contains the following information:

(i) file ownership indication
(ii) file type (e.g., regular, directory, special device, pipes, etc.)
(iii) file access permissions. May have setuid (sticky) bit set
(iv) time of last access, and modification
(v) number of links to the file
(vi) pointers to the data blocks for the file
(vii) size of the file in bytes (for regular files), major and minor device numbers for special devices.

An integral number of inodes fits in a single data block.

Information the inode does not contain:

(i) path (short or full) name of file

Unix stores directories in files. The file type (of the inode) is marked "directory", and the file contains pairs of name/inode numbers.

For example, when a user issues *open("/etc/passwd",...)* the kernel performs the following operations:

1. because the file name is a full path name, find the inode of the root directory (found in superblock) and search the corresponding file for the entry "etc"
2. when the entry "etc" is found, fetch its corresponding inode and check that it is of type directory
3. scan the file associated with "/etc" looking for "passwd"
4. finally, fetch the inode associated with *passwd*'s directory entry, verify that it is a regular file, and start accessing the file.

The Unix file system allocates *data blocks* (blocks that contain a file's contents) one at a time from a pool of free blocks. Unix uses 4K blocks. Moreover, a file's blocks are scattered randomly within the physical disk.

Inodes include pointers to the data blocks. Each inode contains 15 pointers:

(i) the first 12 pointers point directly to data blocks
(ii) the 13th pointer points to an indirect block, a block containing pointers to data blocks
(iii) the 14th pointer points to a doubly-indirect block, a block containing 128 addresses of singly indirect blocks
(iv) the 15th pointer points to a triply indirect block (which contains pointers to doubly indirect blocks, etc.)

```
 -----------------------
 | | | | | | | | | | | | | | | | |
 -----------------------
  | | | | | | | | | | | | | | | | |-----------------------
        Data block         | |-----------|            |
                           |             |            |
                          ----          ----         ----
                          |  |          |  |         |  |
                          ----          ----         ----
                           |||           |||          |||
                          Data          ----         ----
                                        |  |         |  |
                                        ----         ----
                                         |||          |||
                                        Data         ----
                                                     |  |
                                                     ----
                                                      |||
                                                     Data
```

```
with 4K blocks:
direct 12x4K = 48K
indirect 1024x4K = 4MB
double indirect 1024x1024x4K = 4GB
triple indirect 1024x1024x1024x4K = 4TB
```

2.7 ADVANTAGES AND DISADVANTAGES OF UNIX FILE SYSTEM ORGANISATION

1. Advantages

(i) data in small files can be accessed directly from the inode. That is, one read operation fetches the inode, and another read fetches the first data block.

(ii) larger files can be accessed efficiently, because an indirect block points to many data blocks

(iii) disk can be filled completely, with little wasted space (ignoring partially-filled blocks)

2. Disadvantages

(i) because inode information is kept separately from data, access of data often requires a long seek when file is initially accessed

(ii) inodes of files in a common directory not kept together, leading to low performance when searching directories

(iii) original file system only used 512-byte blocks, an inefficient transfer size

(iv) data blocks of a file are not stored together, leading to poor performance when accessing files sequentially

(v) free list quickly becomes scrambled increasing overhead of finding free blocks (seek for each new block)

(vi) original file system used as little as 2% of the available disk bandwidth

☞ SUMMARY

In this chapter, we were introduced to the UNIX system concepts of files and directories, and the basic types of files and directories. We learned about file system, file structure, directory structure etc. We learned how we can list the files in a directory, view the contents of a file. Through UNIX is known to make little distinction the various types of files, that was not really established in this chapter. We used exclusive commands to handle directoreis like pwd, cd, rmdir and mkdir. These commands have no relevance when applied about the type of file it handles.

EXERCISE

1. Describe the file structure of UNIX operating system. What are the differences among director, ordinary file and special files? (UPTU 2001, 03)
2. What is difference between system call and library function? Give some examples for each. (UPTU 2001, 03)
3. Difference between the ordinary files and special files in UNIX operating system. (UPTU 2004)
4. Discuss i-node of UNIX operating system.
5. How are the blocks of file system organized in four categories? Explain each of them.

❑❑❑

CHAPTER 3

Editors (vi and ed)

3.1 INTRODUCTION

In the late 1970's Bill Joy, a graduate student at the University of California at Berkeley, and possibly the main motivator in the development of BSD Unix, created the vi editor.

You may be familiar with some word processors that include hundreds of commands, build-in menus, and features that extend to almost desktop publishing. Well vi is not one of those programs. vi is a visual editor that makes it easy for you to use the full screen as you create and maintain files. Such files may include shell script, source code, mail, documents or reports. One big advantage is that you can edit documents via a telnet window—try doing that with Word!

The basic procedures for using vi are the same for any document. it produces standard ASCII output, which is just what is needed for most purposes.

3.2 vi MODES

When you start up vi, you are automatically placed in command mode. In command mode, you can move the cursor, delete text, change text, perform searches, and exit vi. What you cannot do in command mode is insert text. To do this you need to enter insert mode. This can be achieved by pressing i. When you are finished typing text and want to return to command mode again, you press the escape (ESC) key.

If you are not sure which mode you are in then pressing escape always puts you into command mode. If you are already in command mode, then pressing escape will make your terminal beep, leaves you in command mode any where.

The **:set showmode** option offers an alternative way to distinguish between modes. Setting this option will display a message at the bottom of your screen each time you enter text input mode:

```
– INSERT --
```

There are many flavours of vi, and the one you are using for the tutorials tells you the mode by default without having to type the command shown above.

If you use a set command remember that whatever you set is not saved between runs of vi, so every time you restart vi you must type the set commands again. Set and other ":" commands are entered from command mode by typing a ":" and then the command in question. As you type the command appears at the bottom of the window. Once the command is typed then you press <re-turn> to execute the command.

3.3 GETTING STARTED

In this sample exercise we will be using the following basic commands:

Command	Description
vi test_file	Enter vi and create test_file
i	Insert text before the cursor
ESC	Exit text input mode and return to the command mode
:wq	Write (save) the file and quit vi
:q!	Exit vi without saving changes
l	Move the cursor right
j	Move the cursor down
k	Move the cursor up
h	Move the cursor left
a	Append text after the cursor
x	Delete one character

vi example

1. `% vi  text_file`
2. You should see the following screen with the cursor at the top left.

```
~
~
~
~
~
~
~
~
~
~
"test_file" [New file]
```

3. The "~" character in the window indicates that line currently does not exist. As the file is new no line exists, so this image gives you an idea.

4. What an empty file looks like in vi?
5. Press i (insert before the cursor) to start entering text. This places you in text input mode. The screen does not change and you do not press return after pressing i.
6. Type the following sample text. include the mistakes; you will correct them later. If you type a mistake that is not in the sample, use the backspace key to move back over the incorrect text, and then retype it.
7. As you add text,
9. Will scroll down towards
10. The botom of the screen.
11. Press ESC (escape key). Pressing ESC enters you into command mode, which means you can no longer enter text; instead you can command the cursor to move within the file.
12. Type :wq followed by return. Typing a colon while in command mode opens a command line at the bottom of your screen. When you type the wq command the w writes test_file to the disk, and the q tells vi to quit.
13. Re-enter test_file to correct the mistakes as follows:
14. **$vi test_file**

The command is the same as before, but instead of opening a new test file, vi finds the existing file test_file. The screen clears and you should see the following:

```
As you add text,
will scroll down towards
the botom of the screen.
~
~
~
~
~
~
~
"test_file" 4 lines, 103 characters
```

To correct the two mistakes in the fourth line ("botom" and "screeen"), execute the following:

1. **Press j** three times to move the cursor down to the first character in the forth line.
2. **Press l** (lowercase L) several times to move the cursor right until the cursor is at the 't' in botom. If you pass 't', press h to move the cursor left.
3. **Press i** to insert text before the cursor, and type 't' to spell bottom correctly.
4. **Press ESC** to end the text input mode and return you to the command mode.

Next, correct "screen" which is also in the fourth line:

1. **Press l** until the cursor is at the first e in screen.
2. **Press x** to delete the extra e. If you make a mistake, such as deleting more than one character, press U (uppercase U) to undo the changes you made. After deleting text you do not have to press ESC as you have not left the command mode.

Now, correct the mistake 'wit' in the second line:

1. **Press k** twice to move the cursor up to the second line. Press h to move the cursor left. Position the cursor at the 't' in wit. If you pass the 't', press l to move the cursor right.
2. **Press a** to append text. Pressing a exists command mode, so whatever you type is added to the text.
3. **Type h** to spell 'with' correctly.
4. **Press ESC** to return to the command mode.
5. You now have corrected the mistakes in text_file and can now exit vi in one of two ways:

   :wq

   To exit vi and save the corrections

   :q!

   To exit vi without saving the corrections

3.4 MOVING THE CURSOR

As well as using the keys h,j,k, and l to move the cursor left, down, up, and right respectively, must terminals let you use the arrow keys to perform the same function.

Another useful cursor moving command is G, which functions as a goto command. For example, to go to line 256 of a file you would type 256G and the cursor would move to that line. To find out your current line number at any time, you can press CTRL-G in command mode and to display the line numbers along the left margin of your file, you can type

:set number

$

Moves the cursor to the end of the current line

^ or 0 (zero)

Moves the cursor to the beginning of the current line

w

Moves the cursor to the beginning of the next word or next punctuation

W

Moves the cursor to the beginning of the next word ignoring punctuation

b

Moves the cursor to the beginning of previous word or previous punctuation

B

Moves the cursor to the beginning of previous word ignoring punctuation

e

Move forward to the end of the next word or next punctuation

E

Move forward to the end of the next word ignoring punctuation

3.5 ENTERING TEXT

vi offers various ways to type text in a file. You can insert text before the cursor, append text after the cursor, or open new lines for inserting text above or below the cursor.

i

Insert text before the cursor

I (uppercase i)

Insert text before first character of the line

a attach or add

Append text after the cursor

A

Append text at the end of the line

o

Open a blank line below the cursor for text entry

O

Open a blank line above the cursor for text entry

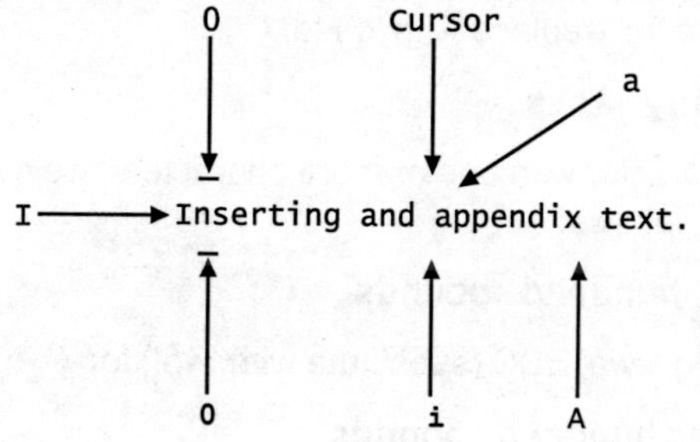

3.6 DELETING TEXT

vi offers various ways to delete text in a file. You can delete text from the cursor back to the first line of text, and from the cursor to the last line of text.

x

Delete character at cursor

X

Delete character after cursor

nx

delete n characters, starting at cursor

dw

deletes one word

ndw

delete from cursor to beginning of nth word

dd

Deletes current line

ndd

Delete next n lines, including the current line

d$

Deletes from cursor to the end of the current line

d^

Deletes from cursor to the beginning of current line

d1G

All lines to beginning of file, including current line

dG

All lines are deleted to the end of the file, including current line.

3.7 REPLACE AND SUBSTITUTE

Within vi you can replace a single character, substitute a single character with one or more characters, and substitute a different word for an existing word. For example, to replace a single characters in command mode, we can place the cursor under the character c in the word ants. So a line:

```
The circus has many ants.
```

Will be changed by typing 'rc' (replace with 'c') to:

```
The circus has many acts.
```

To substitute a single character with one or more characters, we can place the cursor under the character i in the word 'in' as follows:

```
My salary is one hundred pounds.
```

Will be changed by typing 'swa' ESC (substitute with 'wa') to:

```
My salary was one hundred pounds.
```

Notice when you type the s command a dollar sign $ appears in the place of the character. To substitute several characters you can precede the command with the number of characters you want to substitute as follows:

```
My salary is one million pounds.
```

Will be changed by typing '2stri' ESC (substitute two characters with 'tri') to:

```
My salary is one trillion pounds.
```

To change a word, you can use the cw (change word) command, while in command mode, as follows:

```
My salary is one hundred pounds.
```

Will be changed by typing 'cwthousand' ESC to:

```
My salary is one thousand pounds.
```

To substitute several words you could do the following:

```
Change four words in this line
```

Will be changed by typing '4cwEdit' ESC to:

```
Edit this line.
```

cc new words

Replace all characters in a line

C new words

Replace remaining characters in the line from cursor

3.8 MOVING TEXT

To move a block of text from one location to another you can use the delete and put commands. You must be in command mode to achieve this. Suppose you have the following text that you wish to put in the correct order and correct the spelling mistake also. You can do the following:

```
This is the first line of text.
This is the third line of text.
This is the fourth line of text.
This is the second line of text.
```

1. Place the cursor anywhere on the line "This is the second line of text". Type dd to delete the entire line.
2. Move the cursor up and position it anywhere on the first line. Type p to put the deleted text after line one. The text should look as follows.

```
This is the first line of text.
This is the second line of text.
This is the third line of text.
This is the fourth line of text.
```

To correct the transposition error (tetx) in the second line, move the cursor under the 't' in 'tx' and type xp. Typing x deletes the 't'; typing p puts the 't' after the cursor.

p (lowercase p)

Put deleted text below the cursor

P (uppercase p)

Put deleted text above the cursor

xp

Transpose two characters (put deleted character after cursor)

3.9 COPYING TEXT

To copy a few lines of text, you can use the yank command (yy) with the put command.

```
This is the first line of text.
This is the second line of text.
This is the third line of text.
This is the fourth line of text.
```

To duplicate the above lines of text you can proceed as follows:

1. Position the cursor anywhere on the first line.
2. Type the following command to copy the four lines of text: 4yy
3. Move the cursor down to the fourth line.
4. Type p (lower case) to put the copied lines immediately below the fourth linc.

You should now see the following:

```
This is the first  line of text.
This is the second line of text.
This is the third  line of text.
This is the fourth line of text.
This is the first  line of text.
This is the second line of text.
This is the third  line of text.
This is the fourth line of text.
```

yy

Yank (copy) the current line

nyy

Yank the next n lines

yw

Yank a copy of a word

3.10 SCROLLING TEXT

Scrolling moves the cursor backward or forward through the text, so that various lines or sections of the text can be viewed. To execute scroll commands you must be in command mode. Experiment with the following commands using the file "demofile" in your home directory.

CTRL-Y

Scroll backwards one line

*n*CTRL-Y

Scroll backwards n number of line

CTRL-E

Scroll backwards one line

*n*CTRL-E

Scroll forward one line

CTRL-U

Scroll backwards one half screen

CTRL-D

Scroll forwards one half screen line

CTRL-B

Scroll backwards to the previous screen

CTRL-F

Scroll forwards to the next screen

3.11 SEARCHING FOR PATTERNS IN VI

The fastest way to reach a specific place in your text is by using one of the search commands: / (forward slash) or ? (question mark).

To search forward from the current cursor position, use the following command:

```
/pattern
```

where patterns represents the sequence of characters you want to find.

To search backwards from the current cursor position, use the following command:

```
?pattern
```

To search for additional occurrences of the pattern, press either of the following:

n

Repeat the last search command.

N

Repeat the last search command but in the opposite direction.

3.12 Execute file, script or program

When we execute commands at the dollar prompt they are executed in the shell that was invoked when we logged in. When we execute a shell script, the login shell creates a new shell, a new command interpreter and waits, idly in background while the new shell executes our shell script. When all the commands in our shell script have been completed, the new shell terminates and our login shell once again takes over the control.

When we make a script in vi editor then after save the script, we can execute it by using sh command.

For example, suppose we have a shell script like xyz.

```
$ sh xyz ↵
```

3.13 ed EDITOR

The ed command starts the ed editor line-editing program. The ed editor works on only one file at a time by copying it into a temporary edit buffer and making changes to that copy. The ed editor is part of a family of editors that also includes the vi editor, ex editor and edit editor. The ed editor makes the changes you specify in a buffer. It does not alter the file itself until you use the write (w) subcommand.

```
$ed [- p String] [- s | -] [File]
```

You can specify the name of the file you want to edit when you start the ed editor with the ed command, or you can use the e subcommand. When the ed command reads a new file into the buffer, the contents of that file replace the buffer's previous contents.

The red command is a restricted version of the ed command, for use with the restricted shell (rsh). With the red command, you edit only files that reside in the current directory or in the /tmp directory; you cannot use the ! subcommand.

An ed editor subcommand consists of zero, one, or two addresses, followed by a single-character subcommand, followed by optional parameters to that subcommand. The addresses specify one or more lines in the buffer. Because every subcommand has default addresses, it is frequently unnecessary to specify addresses.

The ed editor allows editing only the current line unless you address another line in the buffer. You can move and copy only complete lines of data. The ed editor is useful for editing large files or for editing within a shell program.

The ed editor operates in one of two modes :

command mode	In command mode, the ed editor recognizes and runs subcommands. When you start the ed editor, it is in command mode. Type a . (period) and press the Enter key to confirm that you are in command mode.
text input mode	In text input mode, the ed editor allows you to enter text into the file buffer but does not recognize subcommands. You enter text input mode by using the a subcommand, c subcommand, or i subcommand. You exit text input mode and return to the command mode by typing a . (period) alone at the beginning of a line. To place a .(period) into the buffer while in text input mode, enter a character followed by the .(period). Then, exit text input mode and use the s subcommand to remove the character.

The following list provides the maximum limits of the ed editor :

- 64 characters per file name
- LINE_MAX characters per line (although there is currently a system-imposed limit of 255 characters per line entered from the keyboard)
- 256 characters per global subcommand list
- 128,000 character buffer size

> **Note:** The buffer contains the original file as well as editing information.

The maximum number of lines depends on the amount of memory available. The maximum file size depends on the amount of physical data storage available or on the maximum number of lines permitted in user memory.

3.14 FLAGS

– p String	Sets the editor prompt to the String parameter. The default for String is a null value (no prompt).
– s	Suppresses character counts that the editor displays with the e subcommand, r subcommand, and w subcommand. This flag also suppresses diagnostic messages for the e subcommand and the q subcommand, and suppresses the ! (exclamation point) prompt after an ! subcommand.
–	Provides the same functions as the -s flag.

3.15 PATTERN MATCHING

The ed editor supports a limited form of special pattern-matching characters that you can use as Regular Expressions (REs) to construct pattern strings. You can use these patterns in addresses to specify lines and in some subcommands to specify portions of a line.

3.16. REGULAR EXPRESSIONS

The following Regular Expressions match a single character or a collating element as follows :

Character : Matches itself and can be any ordinary character.

Matches any single character except the new-line character.

[String] : Matches any one character in the string. Certain pattern-matching characters have special meanings within brackets as follows :

^ Matches any character except the characters in the String parameter and the new-line character if the first character of the String parameter is a ^ (circumflex). This condition is true only if the ^ is the first character in the string. For example, ^String.

– Indicates a range of consecutive ASCII characters according to the current collating sequence. For example, [a-f] can be equivalent to [abcdef] or [aAbBcCdDeEfF] or [abcdef] and could even include accented a and e characters. A collating sequence can define equivalence classes for characters.

The minus sign loses its significance if it occurs as the first character in the string, [– String]; if it immediately follows an initial circumflex, [^-String]; or if it appears as the last character in the string, [String –].

] Functions as a part of the string rather than as the string terminator, when the] (right bracket) is the first character in the string, []String], or when it immediately follows an initial circumflex, [^]String].

3.17 FORMING PATTERNS

The following rules describe how to form patterns from Regular Expressions :

(i) A Regular Expression that consists of a single, ordinary character matches that same character in a string.

(ii) A Regular Expression followed by an * (asterisk) matches zero or more occurrences of the character that the RE matches.

For example, the following pattern :

ab*cd

matches each of the following strings :

```
acd
abcd
abbcd
abbbcd
```

but not the following string :

```
abd
```

If a choice exists, the longest matching leftmost string is chosen.

For example, given the following string :

```
122333444
```

the pattern .* matches 122333444, the pattern .*3 matches 122333, and the pattern .*2 matches 122.

(iii) An RE followed by :

\{m\} Matches exactly m occurrences of the character matched by the RE.

\{m,\} Matches at least m occurrences of the character matched by the RE.

\{m,n\} Matches any number of occurrences of the character matched by the RE from m to n inclusive.

(iv) The numbers m and n must be integers from 0 to 255, inclusive. Whenever a choice exists, this pattern matches as many occurrences as possible.

3.18 RESTRICTING: PATTERN MATCHING

You can restrict a pattern to match only the first segment of a line, the final segment, or the entire line. The null pattern, // (two slashes), duplicates the previous pattern.

(i) Matching the First Segment of a Line

The ^Pattern parameter matches only a string that begins in the first character position on a line.

(ii) Matching the Last Segment of a Line

The Pattern$ parameter matches only a string that ends with the last character (not including the new-line character) on a line.

(iii) Matching the Entire Line

The Pattern$ parameter restricts the pattern to match an entire line.

3.19 ADDRESSING LINES

The ed editor uses three types of addresses : line number addresses, addresses relative to the current line, and pattern addresses. The current line (usually the last line affected by a subcommand) is the point of reference in the buffer.

You can use line addressing to do the following :

(i) Designate a new current line

(ii) Display the addressed line or lines

(iii) Cause a command to act on a certain line or lines

Subcommands that do not accept addresses regard the presence of an address as an error. Subcommands that accept addresses can use either given or default addresses. When given more addresses than it accepts, a command uses the last (rightmost) ones.

In most cases, commas (,) separate addresses (for example 2,8). Semicolons (;) also can separate addresses. A semicolon between addresses causes the ed editor to set the current line to the first address and then calculate the second address (for example, to set the starting line for a search). In a pair of addresses, the first address must be numerically smaller than the second.

You can use line numbers and symbolic addresses to perform the following tasks :

1. Addressing the current line
2. Addressing a line by number

3. Addressing the line before the first line
4. Addressing the last line
5. Addressing a line above an addressed line
6. Addressing a line below an addressed line
7. Addressing the first line through the last line
8. Addressing the current line through the last line
9. Addressing a group of lines
10. Addressing the next line that contains a specified pattern
11. Addressing the previous line that contains a specified pattern
12. Addressing a marked line

1. Addressing the Current Line

A . (period) addresses the current line. The . (period) is the default for most ed editor subcommands and does not need to be specified.

2. Addressing a Line by Number

To address a specified line of the buffer, enter :

```
Number$
```

where the Number parameter represents a line number.

For example :

```
2253$
```

addresses line number 2253 as the current line.

3. Addressing the Line before the First Line

To address the line before the first line of the buffer, enter :

```
0
```

4. Addressing the Last Line

To address the last line of the buffer, enter :

```
$
```

5. Addressing a Line above an Addressed Line

To specify an address that is a specified number of lines above the current line, enter :

```
-Number
```

where the Number parameter is the specified number of lines above the current line that you want to address.

For example :

```
-5
```

addresses the line five lines above the current line as the current line.

You also can specify only a - to address the line immediately above the current line. The minus sign has a cumulative effect. For example, the address - - (two minus signs) addresses the line two lines above the current line.

6. Addressing a Line below an Addressed Line

To specify an address that is a specified number of lines below the current line, enter :

```
+Number
```

where the Number parameter is the specified number of lines below the current line that you want to address. The + (plus sign) is optional.

For example :

```
+11
```

addresses the line 11 lines below the current line as the current line.

You also can specify only a + to address the line immediately below the current line. The + has a cumulative effect. For example, the address + + (two plus signs) addresses the line two lines below the current line.

7. Addressing the First Line through the Last Line

To address the first line through the last line, enter :

```
,
```

The , (comma) represents the address pair 1,$ (first line through last line). The first line becomes the current line.

8. Addressing the Current Line through the Last Line

To address the current line through the last line, enter :

```
;
```

The ; (semicolon) represents the address pair .,$ (current line through last line).

9. Addressing a Group of Lines

To address a group of lines, enter :

```
FirstAddress,LastAddress
```

where the FirstAddress parameter is the line number (or symbolic address) of the first line in the group you want to address, and the LastAddress parameter is the line number (or symbolic address) of the last line in the group. The first line in the group becomes the current line. For example :

```
3421,4456
```

addresses the lines 3421 through 4456. Line 3421 becomes the current line.

10. Addressing the Next Line That Contains a Specified Pattern

To address the next line that contains a matching string, enter :

```
/Pattern/
```

where the Pattern parameter is a character string or regular expression. The search begins with the line after the current line and stops when it finds a match for the pattern. If necessary, the search moves to the end of the buffer, wraps around to the beginning of the buffer, and continues until it either finds a match or returns to the current line.

For example :

```
/Tanmay, Arya/
```

addresses the next line that contains Tanmay, Arya as the current line.

11. Addressing the Previous Line That Contains a Specified Pattern

To address the previous line that contains a match for the pattern, enter :

```
?Pattern?
```

where the Pattern parameter is a character string or regular expression. The ?Pattern? construction, like /Pattern/, can search the entire buffer, but it searches in the opposite direction. For example :

```
?Tanmay, Arya?
```

addresses the previous line that contains Tanmay, Arya as the current line.

12. Addressing a Marked Line

To address a marked line with the k subcommand, enter :

```
'x
```

where the x parameter is a lowercase letter a to z. For example :

```
'c
```

addresses the line marked as c with the k subcommand.

3.20 SUBCOMMANDS

Use the ed editor subcommands to perform the following actions :

1. Editing a file
2. Manipulating files
3. Performing miscellaneous functions
    1. Changing the prompt string
    2. Entering system commands
    3. Exiting the ed editor
    4. Requesting help

In most cases, you can enter only one ed editor subcommand on a line. However, you can add the l (list) and p (print) subcommands to any subcommand except the e (edit), E (Edit), f (file), q (quit), Q (Quit), r (read), w (write), and ! (operating system commands) subcommands.

The e, f, r, and w subcommands accept file names as parameters. The ed editor stores the last file name used with a subcommand as a default file name. The next e, E, f, r, or w subcommand given without a file name uses the default file name.

The ed editor responds to an error condition with one of two messages: ? (question mark) or ?File. When the ed editor receives an Interrupt signal (the Ctrl-C key sequence), it displays a ? and returns to command mode. When the ed editor reads a file, it discards ASCII null characters and all characters after the last new-line character.

3.20.1. Editing a File

You can use the ed editor subcommands to perform the following tasks :

1. Adding text
2. Changing text

3. Copying text
4. Deleting text
5. Displaying text
6. Joining and splitting lines
7. Making global changes
8. Marking text
9. Moving text
10. Saving text
11. Searching text
12. Substituting text
13. Undoing text changes

> **Note:** In the following descriptions of ed editor subcommands, default addresses are shown in parentheses. Do not type the parentheses. The address . (period) refers to the current line. A . (period) in the first position of an otherwise empty line is the signal to return to command mode.

1. Adding Text

(.)a [l] [n] [p] Text	. The a (append) subcommand adds text to the buffer after the addressed line. The a subcommand sets the current line to the last inserted line, or, if no lines were inserted, to the addressed line. A 0 address adds text to the beginning of the buffer. Type the l (list), n (number), or p (print) optional subcommand if you want to display the added text. Type your text, pressing the Enter key at the end of each line. If you do not press the Enter key at the end of each line, the ed editor automatically moves your cursor to the next line after you fill a line with characters. The ed editor treats everything you type before you press the Enter key as one line, regardless of how many lines it takes up on the screen. Type a . (period) at the start of a new line, after you have typed all of your text.
(.)i [l] [n] [p]Text.	The i (insert) subcommand inserts text before the addressed line and sets the current line to the last inserted line. If no lines are inserted, the i subcommand sets the current line to the addressed line. You cannot use a 0 address for this subcommand. Type the l (list), n (number), or p (print) optional subcommand if you want to display the inserted text. Type your text, pressing the Enter key at the end of each line. If you do not press the Enter key at the end of each line, the ed editor automatically moves your cursor to the next line after you fill a line with characters. The ed editor treats everything you type before you press the Enter key as one line, regardless of how many lines it takes up on the screen. Type a . (period) at the start of a new line, after you have typed all of your text. Note: The i subcommand differs from the a subcommand only in the placement of the text.

You can use different ed editor subcommands to add text in different locations. Use the preceding format to perform the following editing tasks :

(a) Adding text after the current line
(b) Adding text before the current line
(c) Adding text after an addressed line
(d) Adding text before an addressed line
(e) Adding text after lines that contain a search pattern
(f) Adding text before lines that contain a search pattern
(g) Adding text after lines that do not contain a search pattern
(h) Adding text before lines that do not contain a search pattern

(a) To Add Text after the Current Line

1. Enter the following subcommand :
   ```
   a[l][n][p]
   ```
   where l, n, and p are optional subcommands that display the added text.
2. Type the text, and press the Enter key.
3. Type a . (period), and press the Enter key again to return to command mode.

(b) To Add Text before the Current Line

1. Enter the following subcommand :
   ```
   i[l][n][p]
   ```
   where l, n, and p are optional subcommands that display the added text.
2. Type the text, and press the Enter key.
3. Type a . (period), and press the Enter key again to return to command mode.

(c) To Add Text after an Addressed Line

1. Enter the following subcommand :
   ```
   Addressa[l][n][p]
   ```
   where the Address parameter is the line number of the line that the inserted text should follow. The l, n, and p optional subcommands display the added text.
2. Type the text, and press the Enter key.
3. Type a . (period), and press the Enter key again to return to command mode.

(d) To Add Text before an Addressed Line

1. Enter the following subcommand :
   ```
   Addressi[l][n][p]
   ```
   where the Address parameter is the line number of the line that the inserted text should precede. The l, n, and p optional subcommands display the added text.
2. Type the text, and press the Enter key.
3. Type a . (period), and press the Enter key again to return to command mode.

(e) To Add Text after Lines That Contain a Search Pattern

1. Enter the following subcommand :
   ```
   [Address]g/Pattern/a[l][n][p]
   ```

where Address is an optional parameter that specifies the range of lines to search for the pattern specified in the Pattern parameter. The Pattern parameter is a character string or regular expression. If you omit the Address parameter, the ed editor searches the entire file for lines that contain the pattern. The l, n, and p optional subcommands display the added text.

2. Type a backslash :

   \

3. Type the text. To start new lines within the added text, type a backslash :

   \

   and press the Enter key. The text you type is added after every line that contains the pattern specified in the command.

4. To return to command mode, press the Enter key.

(f) To Add Text before Lines That Contain a Search Pattern

1. Enter the following subcommand :

```
[Address]g/Pattern/i[l][n][p]
```

   where Address is an optional parameter that specifies the range of lines to search for the pattern specified in the Pattern parameter. The Pattern parameter is a character string or regular expression. If you omit the Address parameter, the ed editor searches the entire file for lines that contain the pattern. The l, n, and p optional subcommands display the added text.

2. Type a backslash :

   \

3. Type the text. To start new lines within the added text, type a backslash :

   \

   and press the Enter key. The text you type is added before every line that contains the pattern specified in the command.

4. To return to command mode, press the Enter key.

(g) To Add Text after Lines That Do Not Contain a Search Pattern

1. Enter the following subcommand :

```
[Address]g/Pattern/a[l][n][p]
```

   where Address is an optional parameter that specifies the range of lines to search for lines that do not contain the pattern specified in the Pattern parameter. The Pattern parameter is a character string or regular expression. If you omit the Address, the ed editor searches the entire file for lines that do not contain the pattern. The l, n, and p optional subcommands display the added text.

2. Type a backslash :

   \

3. Type the text. To start new lines within the added text, type a backslash :

   \

   and press the Enter key. The text you type is added after every line that does not contain the pattern specified in the command.

4. To return to command mode, press the Enter key.

(h) To Add Text before Lines That Do Not Contain a Search Pattern

1. Enter the following subcommand :

```
[Address]g/Pattern/i[l][n][p]
```

where Address is an optional parameter that specifies the range of lines to search for lines that do not contain the pattern specified in the Pattern parameter. The Pattern parameter is a character string or regular expression. If you omit the Address parameter, the ed editor searches the entire file for lines that do not contain the pattern. The l, n, and p optional subcommands display the added text.

2. Type a backslash :

\

3. Type the text. To start new lines within the added text, type a backslash :

\

and press the Enter key. The text you type is added before every line that does not contain the pattern specified in the command.

4. To return to command mode, press the Enter key.

2. Changing Text

(.,.)c [l] [n] [p]Text.	The c (change) subcommand deletes the addressed lines you want to replace and then replaces them with the new lines you enter. The c subcommand sets the current line to the last new line of input, or, if no input existed, to the first line that was not deleted. Type the l (list), n (number), or p (print) optional subcommand if you want to display the inserted text. Type the new text, pressing the Enter key at the end of each line. When you have entered all of the new text, type a . (period) on a line by itself.

You can change text in several different ways with the ed editor. Use the preceding format to perform the following editing tasks :

(a) Changing the text of the current line
(b) Changing the text of a line or group of lines
(c) Changing text of lines that contain a specified pattern
(d) Changing text of lines that do not contain a specified pattern

(a) To Change the Text of the Current LIne

1. Enter the following subcommand :

```
c[l][n][p]
```

where l, n, and p are optional subcommands that display the changed text.

2. Type the text, and press the Enter key.
3. Type a . (period), and press the Enter key again to return to command mode.

(b) To Change the Text of a Line or Group of Lines

1. Enter the following subcommand :

```
Addressc[l][n][p]
```

where the Address parameter is the address of the line or group of lines to change. The l, n, and p optional subcommands display the changed text.

2. Type the text, and press the Enter key.
3. Type a . (period), and press the Enter key again to return to command mode.

(c) To Change the Text of Lines That Contain a Specified Pattern

1. Enter the following subcommand :

```
Addressg/Pattern/c[l][n][p]
```

where the Address parameter is the address of the group of lines that you want to search for the pattern specified with the Pattern parameter. The l, n, and p optional subcommands display the changed text.

2. Type a backslash :

```
\
```

3. Type the new text. To start new lines within the new text, type a backslash :

```
\
```

and press the Enter key.

4. To return to command mode, press the Enter key again, type a . (period), and press the Enter key again.

(d) To Change the Text of Lines That Do Not Contain a Specified Pattern

1. Enter the following subcommand :

```
Addressv/Pattern/c[l][n][p]
```

where the Address parameter is the address of the group of lines that you want to search for the pattern specified with the Pattern parameter. The l, n, and p optional subcommands display the changed text.

2. Type a backslash :

```
\
```

3. Type the new text. To start new lines within the new text, type a backslash :

```
\
```

and press the Enter key.

4. To return to command mode, press the Enter key again, type a . (period), and press the Enter key again.

3. Copying Text

(.,.)tAddress [p] [l] [n]	The t (transfer) subcommand inserts a copy of the addressed lines after the line specified by the Address parameter. The t subcommand accepts the 0 address to insert lines at the beginning of the buffer. The t subcommand sets the current line to the last line copied. Type the l (list), n (number), or p (print) optional subcommand if you want to display the transferred text.

Copying a line or a set of lines leaves the specified lines in their original location and puts a copy in the new location. You can select the lines to copy by specifying an address or pattern. Use the preceding format to perform the following editing tasks :

(a) To copying the current line
(b) To copying lines specified by address
(c) To copying lines that contain a specified pattern
(d) To copying lines that do not contain a specified pattern

(a) To Copy the Current Line

1. Enter the following subcommand :

   ```
   tAddress[l][n][p]
   ```

   where the Address parameter is the line number or symbolic address of the line you want a copy of the current line to follow. The l, n, and p optional subcommands display the copied line.
2. Type the text, and press the Enter key.
3. Type a . (period), and press the Enter key again to return to command mode.

(b) To Copy Lines Specified by Address

1. Enter the following subcommand :

   ```
   LineNumbertDestinationAddress[l][n][p]
   ```

   where the LineNumber parameter is the address of the lines you want to copy, and the DestinationAddress parameter is the line you want the copy to follow. The l, n, and p optional subcommands display the copied line.
2. Type the text, and press the Enter key.
3. Type a . (period), and press the Enter key again to return to command mode.

(c) To Copy Lines That Contain a Specified Pattern

Enter the following subcommand :

```
[Address]g/Pattern/t[DestinationAddress][l][n][p]
```

where Address is an optional parameter that specifies the range of lines to search for lines that contain the specified pattern, the Pattern parameter is the text you are searching for, and the DestinationAddress is an optional parameter that identifies the line you want the copied text to follow. The l, n, and p optional subcommands display the copied line.

If you omit the Address parameter, the ed editor searches the entire file for lines that contain the pattern. If you omit the DestinationAddress parameter, the copied text is placed after the current line.

(d) To Copy Lines That Do Not Contain a Specified Pattern

Type the following subcommand :

```
[Address]v/Pattern/t[DestinationAddress][l][n][p]
```

where Address is an optional parameter that specifies the range of lines to search for lines that do not contain the specified pattern, the Pattern parameter is the text, and the DestinationAddress is an optional parameter that identifies the line you want the copied text to follow. The l, n, and p optional subcommands display the copied line.

If you omit the Address parameter, the ed editor searches the entire file for lines that do not contain the pattern. If you omit the DestinationAddress parameter, the copied text is placed after the current line.

4. Deleting Text

(.,.)d [l] [n] [p]	The d (delete) subcommand removes the addressed lines from the buffer. The line after the last line deleted becomes the current line. If the deleted lines were originally at the end of the buffer, the new last line becomes the current line. Type the l (list), n (number), or p (print) optional subcommand if you want to display the deletion.

The ed editor provides several ways to delete text. Use the preceding format to perform the following editing tasks :

(a) Deleting the current line
(b) Deleting a line or group of lines
(c) Deleting a line or group of lines that contain a specified pattern
(d) Deleting a line or group of lines that does not contain a specified pattern
(e) Deleting text from the current line
(f) Deleting text within selected lines
(g) Deleting text from addressed lines
(h) Deleting text from lines that contain a specified pattern
(i) Deleting a pattern from lines that contain a different specified pattern
(j) Deleting a pattern from lines that do not contain a different specified pattern

(a) To Delete the Current Line

Enter the following subcommand :

```
d[l][n][p]
```

where l, n, and p are optional subcommands that display the deleted line.

(b) To Delete a Line or Group of Lines

Enter the following subcommand :

```
Addressd[l][n][p]
```

where the Address parameter is the line number or symbolic address of the lines you want to delete, and l, n, and p are optional subcommands that display the deleted line or lines.

(c) To Delete a Line or Group of Lines That Contain a Specified Pattern

Enter the following subcommand :

```
[Address]g/Pattern/d[l][n][p]
```

where Address is an optional parameter that specifies the line number or symbolic address of the lines you want to search, and the Pattern parameter is a character string or regular expression that represents the text you want to find. If you omit the Address parameter, the ed editor searches the entire file for lines that contain the specified pattern. The l, n, and p optional subcommands display the deleted line or lines.

(d) To Delete a Line or Group of Lines That Does Not Contain a Specified Pattern

Type the following subcommand :

```
[Address]v/Pattern/d[l][n][p]
```

where Address is an optional parameter that specifies the line number or symbolic address of the lines you want to search, and the Pattern parameter is a character string or regular expression that represents the text you want to find. If you omit the Address parameter, the ed editor searches the entire file for lines that do not contain the specified pattern. The l, n, and p optional subcommands display the deleted line or lines.

(e) To Delete Text from the Current Line

1. Type the following subcommand :

   `s/Pattern`

   where the Pattern parameter is a character string or regular expression that represents the text you want to delete.

2. To delete the first instance of the pattern from the line, type :

   `//`

   OR

   To delete every instance of the pattern from the line, type :

   `//g`

3. If you want to display the deletion, type one of the following optional subcommands :

   `l`

   `n`

   `p`

4. Press the Enter key.

(f) To Delete Text within Selected Lines

1. Type the address of a group of lines to select (or skip this step to select all lines).
2. To select the lines indicated by the Pattern parameter in step 4, type :

   `g`

   OR

   To select the lines not indicated by the Pattern parameter in step 4, type :

   `v`

3. To enter the text you want to search, type the following subcommand :

   `/Pattern/s`

   where the Pattern parameter is the text you want to search.
4. Type one of the following commands to make the desired deletion :

   To delete the first instance of the Pattern parameter within each selected line, type :

   `///`

   To delete every instance of the Pattern parameter within each selected line, type :

   `///g`

   To delete the first specified number of occurrences of the Pattern parameter on each selected line (where the Number parameter is an integer), type :

   `///Number`

   To delete the first character string indicated by the OtherPattern parameter within each line selected by the Pattern parameter (where the OtherPattern parameter is the pattern you want to search), type :

```
/OtherPattern//
```

To delete every instance of the OtherPattern parameter within each line selected by the Pattern parameter, type :

```
/OtherPattern//g
```

To delete the first specified number of occurrences of the OtherPattern parameter on each line selected by the Pattern parameter (where the Number parameter is an integer), type :

```
/OtherPattern//Number
```

5. If you want to display the deletion, type one of the following optional subcommands :

```
l
n
p
```

6. Press the Enter key.

For example, to delete all instances of a pattern from a range of lines, type :

```
38,$g/tmp/s/gn
```

The previous example searches all the lines from line 38 to the last line (38,$) for the tmp character string and deletes every instance (/g) of that character string within those lines. It then displays the lines that had text deleted from them and their line numbers (n).

To delete all instances of a pattern from all lines that contain that pattern, type :

```
g/rem/s///gl
```

The previous example searches the entire file (address parameter is omitted) for all lines that contain (g) the rem character string. It deletes all instances (///g) of the rem character string from each of those lines and then displays the lines that had text deleted from them, including the nonprinting characters in those lines (l).

(g) To Delete Text from Addressed Lines

1. Type the following subcommand :

   Note : The Address parameter is followed by the s subcommand.

   ```
   Addresss/Pattern
   ```

   where the Address parameter is the line number, range of line numbers, or symbolic address of the lines you want to delete the pattern from, and the Pattern parameter is a character string or regular expression that represents the text you want to delete.

2. To delete the first instance of the pattern from each line, type :

   ```
   //
   ```

   OR

   To delete every instance of the pattern from each line, type :

   ```
   //g
   ```

3. If you want to display the deletion, type one of the following optional subcommands :

   ```
   l
   n
   p
   ```

4. Press the Enter key.

(h) To Delete Text from Lines That Contain a Specified Pattern

1. Type the following subcommand:

   ```
   [Address]g/Pattern/s
   ```

   where Address is an optional parameter that specifies the line number, range of line numbers, or symbolic address of the lines that contains a specified pattern, and the Pattern parameter is a character string or regular expression that represents the text you want to find and delete. If you omit the Address parameter, the ed editor searches all lines in the file for the pattern.

2. To delete the first instance of the pattern from each line that contains it, type :

   ```
   ///
   ```

   OR

   To delete every instance of the pattern from each line that contains it, type :

   ```
   ///g
   ```

3. If you want to display the deletion, type one of the following optional subcommands :

   ```
   l
   n
   p
   ```

4. Press the Enter key.

(i) To Delete a Pattern from Lines That Contain a Different Specified Pattern

1. Type the following subcommand :

   ```
   [Address]g/SearchPattern/s
   ```

   where Address is an optional parameter that specifies the line number, range of line numbers, or symbolic address of the lines that contains a specified pattern, and the SearchPattern parameter is a character string or regular expression that represents text that is in the lines you want to change. If you omit the Address parameter, the ed editor searches all lines in the file for the specified pattern.

2. To specify the text you want to delete, type :

   ```
   /DeletePattern/
   ```

3. To delete the first instance of the pattern from each line, type :

   ```
   /
   ```

   OR

   To delete every instance of the pattern from each line, type :

   ```
   /g
   ```

   Note : The entire subcommand string looks like this :

   ```
   [Address]g/SearchPattern/s/DeletePattern//[g]
   ```

4. If you want to display the deletion, type one of the following optional subcommands :

   ```
   l
   n
   p
   ```

5. Press the Enter key.

For example, to delete the first instance of a pattern from lines that contain a different specified pattern, type :

```
1,.g/rem/s/tmp//l
```

The previous example searches from the first line to the current line (1,.) for all lines that contain (g) the rem character string. It deletes the first instance of the tmp character string from each of those lines (/), then displays the lines that had text deleted from them, including the nonprinting characters in those lines (l).

(j) To Delete a Pattern from Lines That Do Not Contain a Different Specified Pattern

1. Type the following subcommand :

   ```
   [Address]v/SearchPattern/s
   ```

   where Address is an optional parameter that specifies the line number, range of line numbers, or symbolic address of the lines that contains a specified pattern, and the SearchPattern parameter is a character string or regular expression that represents text that is not in the lines you want to find and change. If you omit the Address parameter, the ed editor searches all lines in the file for the specified pattern.

2. To specify the text you want to delete, type :

   ```
   /DeletePattern/
   ```

3. To delete the first instance of the pattern, type :

   ```
   /
   ```

   OR

   To delete every instance of the pattern from each line, type :

   ```
   /g
   ```

   Note : The entire subcommand string looks like this :

   ```
   [Address]v/SearchPattern/s/DeletePattern//[g]
   ```

4. If you want to display the deletion, type one of the following optional subcommands :

   ```
   l
   n
   p
   ```

5. Press the Enter key.

For example, to delete the first instance of a pattern from lines that do not contain a specified pattern, type :

```
1,.v/rem/s/tmp//l
```

The previous example searches from the first line to the current line (1,.) for all lines that do not contain (v) the rem character string. It deletes the first instance of the tmp character string from each of those lines (/), then displays the lines that had text deleted from them, including the nonprinting characters in those lines (l).

5. Displaying Text

(.,.)l	The l (list) subcommand writes the addressed lines to standard output in a visually unambiguous form and writes the characters \\\, \\a, \\b, \\f, \\r, \\t, and \\v in the corresponding escape sequence. The lsubcommand writes nonprintable

	characters as one 3-digit octal number, with a preceding \ (backslash) for each byte in the character (most significant byte first). The l subcommand wraps long lines, and you can indicate the wrap point by writing the \ (backslash)/new-line character sequence. Wrapping occurs at the 72nd column position. The $ (dollar sign) marks the end of each line. You can append the l subcommand to any ed editor subcommand except the e, E, f, q, Q, r, w, or ! subcommand. The current line number is set to the address of the last line written.
(.,.)n	The n (number) subcommand displays the addressed lines, each preceded by its line number and a tab character (displayed as blank spaces); n sets the current line to the last line displayed. You can append the n subcommand to any ed editor subcommand except e, f, r, or w. For example, the dn subcommand deletes the current line and displays the new current line and line number.
(.,.)p	The p (print) subcommand displays the addressed lines and sets the current line to the last line displayed. You can append the p subcommand to any ed editor subcommand except e, f, r, or w. For example, the dp subcommand deletes the current line and displays the new current line.
(.)=	Without an address, the = (equal sign) subcommand displays the current line number. When preceded by the $ address, the = subcommand displays the number of the last line in the buffer. The = subcommand does not change the current line and cannot be appended to a g subcommand or v subcommand.

When you search for lines that contain or do not contain a specified pattern, you can select a range of line numbers to search. You can select and display one line or a group of lines in an ed editor file several different ways. Use the preceding format to perform the following editing tasks :

(a) Displaying an addressed line or group of lines
(b) Displaying an addressed line or group of lines and their nonprinting characters
(c) Displaying an addressed line or group of lines and their line numbers
(d) Displaying lines that contain a search pattern
(e) Displaying lines that contain a search pattern, including their nonprinting characters
(f) Displaying lines that contain a search pattern, including their line numbers
(g) Displaying lines that do not contain a search pattern
(h) Displaying lines that do not contain a search pattern, including their nonprinting characters
(i) Displaying lines that do not contain a search pattern, including their line numbers

(a) To Display an Addressed Line or Group of Lines

Enter the following subcommand :

```
Addressp
```

where the Address parameter is the line number or symbolic address of the lines you want to display.

The line or lines addressed are displayed on the screen. If the group of lines is too long to fit on the screen, the ed editor displays as many as will fit, beginning with the first line addressed.

(b) To Display an Addressed Line or Group of Lines and Their Nonprinting Characters

Enter the following subcommand :

```
Addressl
```

where the Address parameter is the line number or symbolic address of the lines you want to display.

The line or lines addressed and their non-printing characters are displayed on the screen. If the group of lines is too long to fit on the screen, the ed editor displays as many as will fit, beginning with the first line addressed.

(c) To Display an Addressed Line or Group of Lines and Their Line Numbers

Enter the following subcommand :

```
Addressn
```

where the Address parameter is the line number or symbolic address of the lines you want to display.

The line or lines addressed are displayed on the screen. The line number for each line is displayed beside the line. If the group of lines is too long to fit on the screen, the ed editor displays as many as will fit, beginning with the first line addressed.

(d) To Display Lines That Contain a Search Pattern

Enter the following subcommand :

```
Addressg/Pattern/p
```

where the Address parameter is the range of lines and the Pattern parameter is the character string or regular expression that you want to search.

The line or lines that contain the specified pattern are displayed on the screen. If the group of lines is too long to fit on the screen, the ed editor displays as many as will fit, beginning with the first line addressed.

(e) To Display Lines That Contain a Search Pattern, Including Their Non-printing Characters

Enter the following subcommand :

```
[Address]g/Pattern/l
```

where Address is an optional parameter that specifies the range of lines and the Pattern parameter is the character string or regular expression that you want to search. If you omit the Address parameter, the ed editor searches the entire file.

The line or lines that contain the specified pattern are displayed on the screen. Non-printing characters show up in the display. If the group of lines is too long to fit on the screen, the ed editor displays as many as will fit, beginning with the first line addressed.

(f) To Display Lines That Contain a Search Pattern, Including Their Line Numbers

Enter the following subcommand :

```
[Address]g/Pattern/n
```

where Address is an optional parameter that specifies the range of lines and the Pattern parameter is the character string or regular expression that you want to search. If you omit the Address parameter, the ed editor searches the entire file.

The line or lines that contain the specified pattern are displayed on the screen. The line number for each line is displayed beside the line. If the group of lines is too long to fit on the screen, the ed editor displays as many as will fit, beginning with the first line addressed.

(g) To Display Lines That Do Not Contain a Search Pattern

Enter the following subcommand :

```
[Address]v/Pattern/p
```

where Address is an optional parameter that specifies the range of lines and the Pattern parameter is the character string or regular expression that you want to search. If you omit the Address parameter, the ed editor searches the entire file.

The line or lines that do not contain the specified pattern are displayed on the screen. If the group of lines is too long to fit on the screen, the ed editor displays as many as will fit, beginning with the first line addressed.

(i) To Display Lines That Do Not Contain a Search Pattern, Including Their Non-printing Characters

Enter the following subcommand :

```
[Address]v/Pattern/l
```

where Address is an optional parameter that specifies the range of lines and the Pattern parameter is the character string or regular expression that you want to search. If you omit the Address parameter, the ed editor searches the entire file.

The line or lines that do not contain the specified pattern are displayed on the screen, including the non-printing characters. If the group of lines is too long to fit on the screen, the ed editor displays as many as will fit, beginning with the first line addressed.

(j) To Display Lines That Do Not Contain a Search Pattern, Including Their Line Numbers

Enter the following subcommand :

```
[Address]v/Pattern/n
```

where Address is an optional parameter that specifies the range of lines and the Pattern parameter is the character string or regular expression that you want to search. If you omit the Address parameter, the ed editor searches the entire file.

The line or lines that do not contain the specified pattern are displayed on the screen, along with their line numbers. If the group of lines is too long to fit on the screen, the ed editor displays as many as will fit, beginning with the first line addressed.

6. Joining and Splitting Lines

(.,.+1)j [l] [n] [p]	The j (join) subcommand joins contiguous lines by removing the intervening new-line characters. If given only one address, the j subcommand does nothing. Type the l (list), n (number), or p (print) subcommand if you want to display the joined lines. These subcommands are optional.

The ed editor provides several ways to join or split a line. Use the preceding format to perform the following editing tasks :

(a) To joining the current and next lines

(b) To joining addressed lines

(c) To splitting the current line

(d) To splitting an addressed line

(a) To Join the Current and Next Lines

Enter the following subcommand :

```
j[l][n][p]
```

where l, n, and p are optional subcommands that display the joined lines.

(b) To Join Addressed Lines

Enter the following subcommand :

```
Addressj[l][n][p]
```

where the Address parameter is a set of contiguous lines that will form one line, and l, n, and p are optional subcommands that display the joined lines.

(c) To Split the Current Line

1. To split the current line after a specified pattern, type the following subcommand :

   ```
   s/Pattern/Pattern\
   ```

   where the Pattern parameter is the character string that you want to split the line after.

   Note : Make sure that both strings represented by the Pattern parameter are exactly alike.
2. Press the Enter key.
3. Type the following backslash :

   ```
   /
   ```
4. To display the split line, type one of the following optional subcommands :

   ```
   l
   n
   p
   ```
5. Press the Enter key.

(d) To Split an Addressed Line

1. To split an addressed line after a specified pattern, type the following subcommand :

   ```
   Addresss/Pattern/Pattern\
   ```

   where the Address parameter is the address of the line to split, and the Pattern parameter is the character string to split the line after.

   Note : Make sure that both strings represented by the Pattern parameter are exactly alike.
2. Press the Enter key.
3. Type the following backslash :

   ```
   /
   ```
4. To display the split line, type one of the following optional subcommands :

   ```
   l
   n
   p
   ```
5. Press the Enter key.

7. Making Global Changes

(1,$)g/Pattern/SubcommandList [l] [n] [p]	The g (global) subcommand first marks every line that matches the Pattern parameter. The pattern can be a fixed character string or a regular

	expression. Then, for each marked line, this subcommand sets the current line to the marked line and runs the SubcommandList parameter. Enter a single subcommand or the first subcommand of a list of subcommands on the same line with the g subcommand; enter subsequent subcommands on separate lines. Except for the last line, each of the lines should end with a \ (backslash). The SubcommandList parameter can include the a, i, and c subcommands and their input. If the last command in the SubcommandList parameter would normally be the . (period) that ends input mode, the . (period) is optional. If no SubcommandList parameter exists, the current line is displayed. The SubcommandList parameter cannot include the g , G, v, or V subcommand. Type the I (list), n (number), or p (print) subcommand if you want to display the changes. These subcommands are optional. Note: The g subcommand is similar to the v subcommand, which runs the SubcommandList parameter for every line that does not contain a match for the pattern.
(1,$)G/Pattern/ [l] [n] [p]	The interactive G (Global) subcommand marks every line that matches the Pattern parameter, displays the first marked line, sets the current line to that line, and then waits for a subcommand. A pattern can be a fixed character string or a regular expression. The G subcommand does not accept the a, i, c, g, G, v, and V subcommands. After the subcommand finishes, the G subcommand displays the next marked line, and so on. The G subcommand takes a new-line character as a null subcommand. A :& (colon ampersand) causes the G subcommand to run the previous subcommand again. You can stop the G subcommand by typing the Ctrl-C key sequence. Type the I (list), n (number), or p (print) subcommand if you want to display the changes. These subcommands are optional.
(1,$)v/Pattern/SubcommandList [l] [n] [p]	The v subcommand runs the subcommands in the SubcommandList parameter for each line that does not contain a match for the Pattern parameter. A pattern can be a fixed character string or a regular expression. Type the I (list), n

	(number), or p (print) subcommand if you want to display the changes. These subcommands are optional. The v subcommand does not accept the a, i, c, g, G, and V subcommands. Note: The v subcommand complements the g subcommand, which runs the SubcommandList parameter for every line that contains a match for the pattern.
(1,$)V/Pattern/ [l] [n] [p]	The V subcommand marks every line that does not match the Pattern parameter, displays the first marked line, sets the current line to that line, and then waits for a subcommand. A pattern can be a fixed character string or a regular expression. Type the l (list), n (number), or p (print) subcommand if you want to display the changes. These subcommands are optional. The V subcommand does not accept the a, i, c, g, G, and v subcommands. Note: The V subcommand complements the G subcommand, which marks the lines that match the pattern.

8. Marking Text

(.)kx [l] [n] [p]	The k (mark) subcommand marks the addressed line with the name specified by the x parameter, which must be a lowercase ASCII letter. The address 'x (single quotation mark before the marking character) then addresses this line. The k subcommand does not change the current line. Type the l (list), n (number), or p (print) subcommand if you want to display the marked text. These subcommands are optional.

(a) To Mark the Current Line

Enter the following subcommand :

```
kLetter[l][n][p]
```

where the Letter parameter is the letter a through z for a mark, and l, n, and p are optional subcommands that display the marked text.

(b) To Mark an Addressed Line

Enter the following subcommand :

```
AddresskLetter[l][n][p]
```

where the Address parameter is the line number or symbolic address of the line you want to mark, and the Letter parameter is the letter a through z for a mark. The l, n, and p optional subcommands display the marked text.

9. Moving Text

(.,.)mA [l] [n] [p]	The m (move) subcommand repositions the addressed line or lines. The first moved line follows the line addressed by the A parameter. A parameter of 0 moves the addressed line or lines to the beginning of the file. The

	address specified by the A parameter cannot be one of the lines to be moved. The m subcommand sets the current line to the last moved line. Type the l (list), n (number), or p (print) subcommands if you want to display the deletion. These subcommands are optional.

Moving a line or a set of lines deletes the specified lines from their original location and places them in a new location. You can select which lines to move by address or pattern. Use the preceding format to perform the following editing tasks :

(a) To move the current line

(b) To move lines specified by address

(c) To move lines that contain a specified pattern

(d) To move lines that do not contain a specified pattern

(a) To Move the Current Line

Enter the following subcommand :

```
mAddress[l][n][p]
```

where the Address parameter is the line number or symbolic address of the line you want the current line to follow, and l, n, and p are optional subcommands that display the moved line.

(b) To Move Lines Specified by Address

Enter the following subcommand :

```
LineNumbermDestinationAddress[l][n][p]
```

where the LineNumber parameter is the address of the lines you want to move, and the DestinationAddress parameter is the line you want the moved lines to follow. The l, n, and p optional subcommands display the moved lines.

(c) To Move Lines That Contain a Specified Pattern

Enter the following subcommand :

```
[Address]g/Pattern/m[DestinationAddress][l][n][p]
```

where Address is an optional parameter that specifies the range of lines to search for lines that contain the specified pattern, the Pattern parameter is the text you are searching for, and DestinationAddress is an optional parameter that represents the line you want the moved lines to follow. The l, n, and p optional subcommands display the moved lines.

If you omit the Address parameter, the ed editor searches the entire file for lines that contain the pattern. If you omit the DestinationAddress parameter, the moved text is placed after the current line.

(d) To Move Lines That Do Not Contain a Specified Pattern

Enter the following subcommand :

```
[Address]v/Pattern/m[DestinationAddress][l][n][p]
```

where Address is an optional parameter that specifies the range of lines to search for lines that do not contain the specified pattern, the Pattern parameter is the text, and DestinationAddress is an optional parameter that represents the line you want the moved text to follow. The l, n, and p optional subcommands display the moved lines.

If you omit the Address parameter, the ed editor searches the entire file for lines that do not contain the pattern. If you omit the DestinationAddress parameter, the moved text is placed after the current line.

10. Saving Text

(1,$)w File	The w (write) subcommand copies the addressed lines from the buffer to the file specified by the File parameter. If the file does not exist, the w subcommand creates it with permission code 666 (read and write permission for everyone), unless the umask setting specifies another file creation mode.
	The w subcommand does not change the default file name (unless the File parameter is the first file name used since you started the ed editor). If you do not provide a file name, the w subcommand uses the default file name. The w subcommand does not change the current line.
	If the ed editor successfully writes the file from the buffer, it displays the number of characters written. If you specify the ! Command subcommand instead of a file name, the w subcommand reads the output of the operating system command specified by the Command parameter. The w subcommand does not save the name of the operating system command you specified as a default file name.
	Note : Because 0 is not a legal address for the w subcommand, you cannot create an empty file with the ed command.

You can save changes to a file in several ways. Use the preceding format to perform the following actions :

(a) To saving a file to the current file
(b) To saving part of a file to the current file
(c) To saving a file to a different file
(d) To saving part of a file to a different file

(a) To Save a File to the Current File

Enter the following subcommand :

```
w
```

The current file is saved under its current name, and the ed editor displays the number of characters written.

(b) To Save Part of a File to the Current File

Enter the following subcommand :

```
Addressw
```

where the Address parameter specifies the line or group of lines to write. The ed editor displays the number of characters written.

(c) To Save a File to a Different File

Enter the following subcommand :

```
w File
```

where the File parameter is the name of the file to write to.

The current file is saved to the file specified by the File parameter. The ed editor displays the number of characters written.

(d) To Save Part of a File to a Different File

Enter the following subcommand :

```
Addressw File
```

where the Address parameter specifies the line or group of lines to write and the File parameter specifies the file to write to.

The specified lines are saved to the file specified by the File parameter. The ed editor displays the number of characters written.

11. Searching Text

You can search forward or backward from the current line for a pattern of text. The pattern can be a character string or a regular expression made up of literal characters and the special characters ^ (circumflex), $ (dollar sign), . (period), [(left bracket),] (right bracket), * (asterisk), \ (reverse slash), & (ampersand), and % (percent sign).

You can use the ed editor to perform the following text searches :

(a) To searching forward

(b) To searching backward

(c) To repeating a search in the same direction

(d) To repeating a search in the opposite direction

(a) To Search Forward

Enter the following subcommand :

```
/Pattern
```

where the Pattern parameter is a character string or regular expression that specifies the text to search for.

The cursor moves to the first character of the text specified by the pattern.

(b) To Search Backward

Enter the following subcommand :

```
?Pattern
```

where the Pattern parameter is a character string or regular expression that specifies the text to search for.

The cursor moves to the first character of the text specified by the pattern.

(c) To Repeat a Search in the Same Direction

Enter the following subcommand :

```
/
```

The cursor moves to the first character of the closest instance of the text specified by the pattern in the last search command.

(d) To Repeat a Search in the Opposite Direction

Enter the following subcommand :

```
?
```

The cursor moves to the first character of the closest instance of the text specified by the pattern in the last search command.

12. Substituting Text

(.,.)s/Pattern/Replacement/ [l] [n] [p] (.,.)s/Pattern/Replacement/ng [l] [n] [p]	The s (substitute) subcommand searches each addressed line for a string that matches the Pattern parameter and replaces the string with the specified Replacement parameter. A pattern can be a fixed character string or a regular expression. Without the global subcommand (g), the s subcommand replaces only the first matching string on each addressed line. With the g subcommand, the s subcommand replaces every occurrence of the matching string on each addressed line. If the s subcommand does not find a match for the pattern, it returns the error message ? (question mark).
	Type the l (list), n (number), or p (print) subcommand to display the substituted text. These subcommands are optional.
	Note : Any character except a space or a new-line character can separate (delimit) the Pattern and Replacement parameters. The s subcommand sets the current line to the last line changed.
	If the Number parameter (an integer) is specified, then the first number that matches strings in each addressed line is replaced.
	An & (ampersand) character used in the Replacement parameter has the same value as the Pattern parameter. For example, the subcommand s/are/&n't/ has the same effect as the subcommand s/are/aren't/ and replaces are with aren't on the current line. A \& (backslash, ampersand) removes the special meaning of the & (ampersand) in the Replacement parameter.
	A subpattern is part of a pattern enclosed by the strings \((backslash, left parenthesis) and \) (backslash, right parenthesis); the pattern works as if the enclosing characters were not present. In the Replacement parameter, \Number refers to strings that match subpatterns. For example, the s/\(t\)\(h\) \(e\)/t\1\2ose) subcommand replaces the with those if a match for the pattern the exists on the current line. Whether subpatterns are nested or in a series, \Number refers to the occurrence specified by the Number parameter, counting from the left of the delimiting characters, \) (backslash, right parenthesis).

	The % (percent sign), when used alone as the Replacement parameter, causes the s subcommand to repeat the previous Replacement parameter. The % does not have this special meaning if it is part of a longer Replacement parameter or if it is preceded by a \ (backslash).
	You can split lines by substituting new-line characters into them. In the Replacement parameter, the \-Enter key sequence quotes the new-line character (not displayed) and moves the cursor to the next line for the remainder of the string. New-line characters cannot be substituted as part of a g subcommand or v subcommand list.

The ed editor provides several ways to substitute text. Use the preceding format to perform the following editing tasks :

(a) To substituting text within the current line

(b) To substituting text within an addressed line or group of lines

(c) To substituting a specified pattern within lines that contain that pattern

(d) To substituting a pattern within lines that contain a different pattern

(e) To substituting a pattern within lines that do not contain a different pattern

(a) To Substitute Text within the Current Line

1. Type the following subcommand :

   ```
   s/OldString/NewString
   ```

   where the OldString parameter is the existing text and the NewString parameter is the text you want to substitute for it.

2. Type one of the following actions :

   To substitute the NewString parameter for the first instance of the OldString parameter within the current line, type :

   ```
   /
   ```

   To substitute the NewString parameter for every instance of the OldPattern parameter within the current line, type :

   ```
   /g
   ```

3. To display the changed text, type one of the following optional subcommands :

   ```
   l
   n
   p
   ```

4. Press the Enter key.

(b) To Substitute Text within an Addressed Line or Group of Lines

1. Type the following subcommand :

   ```
   Addresss/OldPattern/NewString
   ```

where the Address parameter is the address of the line or group of lines where you want to substitute text, the OldPattern parameter is the existing text, and the NewString parameter is the text you want to substitute.

2. Type one of the following actions :

   To substitute the NewString parameter for the first instance of the OldPattern parameter within each line, type :

   `/NewString/`

   To substitute the NewString parameter for every instance of the OldPattern parameter within each line, type :

   `/NewString/g`

   To substitute the NewString parameter for the first instance of the NumberOldPattern parameter on each address line, type :

   `/NewString/Number`

3. To display the changed text, type one of the following optional subcommands :

   `l`

   `n`

   `p`

4. Press the Enter key.

(c) To Substitute a Specified Pattern within Lines That Contain That Pattern

1. Type the following subcommand :

   `Addressg/Pattern/s//NewString`

   where the Address parameter is the address of the group of lines that you want to search for the pattern specified with the Pattern parameter, and the NewString parameter is the text you want to substitute for the Pattern parameter.

2. Type one of the following actions :

   To substitute the NewString parameter for the first instance of the Pattern parameter within each line, type :

   `/`

   To substitute the NewString parameter for every instance of the Pattern parameter within each line, type :

   `/g`

3. To display the changed text, type one of the following optional subcommands :

   `l`

   `n`

   `p`

4. Press the Enter key.

(d) To Substitute a Pattern within Lines That Contain a Different Pattern

1. Type the following subcommand :

   `Addressg/Pattern/s/OldString/NewString`

where the Address parameter is the address of the group of lines that you want to search for the pattern specified with the Pattern parameter, the OldString parameter is the text you want to replace, and the NewString parameter is the text you want to substitute in place of the OldString parameter.

2. Type one of the following actions :

   To substitute the NewString parameter for the first instance of the OldString parameter within each line that contains the Pattern parameter, type :

   ```
   /
   ```

   To substitute the NewString parameter for every instance of the OldString parameter within each line that contains the Pattern parameter, type :

   ```
   /g
   ```

3. To display the changed text, type one of the following optional subcommands :

   ```
   l
   n
   p
   ```

4. Press the Enter key.

(e) To Substitute a Pattern within Lines That Do Not Contain a Different Pattern

1. Type the following subcommand :

   ```
   Addressv/Pattern/s/OldString/NewString
   ```

   where the Address parameter is the address of the group of lines that you want to search for the pattern specified with the Pattern parameter, the OldString parameter is the text you want to replace, and the NewString parameter is the text you want to substitute in place of the OldString parameter.

2. Type one of the following actions :

   To substitute the NewString parameter for the first instance of the OldString parameter within each line that does not contain the Pattern parameter, type :

   ```
   /
   ```

   To substitute the NewString parameter for every instance of the OldString parameter within each line that does not contain the Pattern parameter, type :

   ```
   /g
   ```

3. To display the changed text, type one of the following optional subcommands :

   ```
   l
   n
   p
   ```

4. Press the Enter key.

13. Undoing Text Changes

u [l] [n] [p]	The u (undo) subcommand restores the buffer to the state it was in before it was last modified by an ed editor subcommand. The u subcommand cannot undo the e, f, and w subcommands. Type the l (list), n (number), or p (print) subcommand if you want to display the changes. These subcommands are optional.

(a) To Undo Text Changes

Enter the following subcommand :

```
u[l][n][p]
```

where l, n, and p are optional subcommands that display the changes. All add, change, move, copy, or delete editing functions performed to the text after the last save are undone.

3.20.2 Manipulating Files

You can use ed editor subcommands to manipulate files to perform the following tasks :

1. Adding another file to the current file
2. Changing the default file name
3. Editing additional files

1. Adding Another File to the Current File

($)r File	The r (read) subcommand reads a file into the buffer after the addressed line. The r subcommand does not delete the previous contents of the buffer. When entered without the File parameter, the r subcommand reads the default file, if any, into the buffer. The r subcommand does not change the default file name.
	A 0 address causes the r subcommand to read a file in at the beginning of the buffer. After it reads a file successfully, the r subcommand displays the number of characters read into the buffer and sets the current line to the last line read.
	If ! (exclamation point) replaces the File parameter in an r subcommand, the rest of the line is taken as an operating system shell command whose output is to be read. The r subcommand does not store the names of operating system commands as default file names.

(a) To Insert a File after the Current Line

Enter the following subcommand :

```
r File
```

where the File parameter is the name of the file to be inserted.

The ed editor reads the file specified by the File parameter into the current file after the current line and displays the number of characters read into the current file.

(b) To Insert a File after a Line Specified by Address

Enter the following subcommand :

```
Addressr File
```

where the Address parameter specifies the line that you want the inserted file to follow, and the File parameter is the name of the file to be inserted.

The ed editor reads the file specified by the File parameter into the current file after the specified line and displays the number of characters read into the current file.

2. Changing the Default File Name

f [File]	The f (file name) subcommand changes the default file name (the stored name of the last file used) to the name specified by the File parameter. If a File parameter is not specified, the f subcommand displays the default file name. (The e subcommand stores the default file name.)

(a) To Display the Name of a File

Enter the following subcommand :

```
f
```

The ed editor displays the name of the file in the edit buffer.

(b) To Name a File

Enter the following subcommand :

```
f File
```

where the File parameter is the new name for the file in the edit buffer.

The file in the edit buffer is renamed.

3. Editing Additional Files

e File	The e (edit) subcommand first deletes any contents from the buffer, sets the current line to the last line of the buffer, and displays the number of characters read into the buffer. If the buffer has been changed since its contents were saved (with the w subcommand), the ed editor displays a ? (question mark) before it clears the buffer. The e subcommand stores the File parameter as the default file name to be used, if necessary, by subsequent e, r, or w subcommands. (To change the name of the default file name, use the f subcommand.)
	When an ! (exclamation point) replaces the File parameter, the e subcommand takes the rest of the line as an operating system shell command and reads the command output. The e subcommand does not store the name of the shell command as a default file name.
E File	The E (Edit) subcommand works like the e subcommand with one exception; the E subcommand does not check for changes made to the buffer after the last w subcommand. Any changes you made before re-editing the file are lost.

You can use the e or E subcommands to perform the following tasks :

(a) To re-editing the current file without saving it

(b) To re-editing the current file after saving it

(c) To editing a file after the current file Is saved

(d) To editing a file without saving the current file

(a) To Re-Edit the Current File without Saving It

Enter the following subcommands :

```
E
```

The ed editor displays the number of characters in the file. Any changes you made before re-editing the file are lost.

(b) To Re-Edit the Current File after Saving It

Enter the following subcommands :

```
e
```

The ed editor displays the number of characters in the file.

(c) To Edit a File after the Current File Is Saved

Enter the following subcommands :

```
e File
```

where the File parameter is the name of a new or existing file that you want to edit.

For an existing file, the ed editor displays the number of characters in the file. For a new file, the ed editor displays a ? (question mark) and the name of the file.

(d) To Edit a File without Saving the Current File

Enter the following subcommands :

```
E File
```

where the File parameter is the name of a new or existing file that you want to edit.

For an existing file, the editor displays the number of characters in the file. For a new file, the ed editor displays a ? (question mark) and the name of the file.

3.20.3. Miscellaneous Functions of the ed Editor Subcommands

You can use ed editor subcommands to perform the following tasks :

1. Changing the prompt string
2. Entering system commands
3. Exiting the ed editor
4. Requesting help

1. the Prompt String

P	The P (Prompt) subcommand turns on or off the ed editor prompt string, which is represented by an * (asterisk). Initially, the P subcommand is turned off.

To Start or Stop Displaying the Prompt String

Enter the following subcommand :

```
P
```

The ed editor prompt, an * (asterisk), is displayed or not displayed, depending on its previous setting.

2. Entering System Commands

! Command	The ! subcommand allows you to run operating system commands without leaving the ed editor. Anything that follows the ! subcommand on an ed

	editor subcommand line is interpreted as an operating system command. Within the text of that command string, the ed editor replaces the unescaped % (percent sign) with the current file name, if one exists.
	You can repeat the previous operating system command by entering an ! (exclamation point) after the ! ed editor subcommand. If the operating system command interpreter (the sh command) expands the command string, the ed editor echoes the expanded line. The ! subcommand does not change the current line.

You can use the ! subcommand to perform the following actions :

(a) To running one operating system command

(b) To repeating an operating system command

(c) To running several operating system commands

(a) To Run One Operating System Command

Enter the following subcommand :

```
!Command
```

where the Command parameter specifies an operating system command usually entered at the prompt.

The command runs and displays its output. After the command completes, the editor displays an ! (exclamation point).

(b) To Repeat an Operating System Command

Enter the following subcommand :

```
!
```

The previously run operating system command runs and displays its output. After the command completes, the editor displays an ! (exclamation point).

(c) To Run Several Operating System Commands

(i) Type the following subcommand to display an operating system prompt:

```
!sh
```

(ii) Type an operating system command.

(iii) Press the Enter key to run the command and display its output.

(iv) Repeat steps 2 and 3 to run more operating system commands.

(v) Press the Ctrl-D key sequence to return to command mode. The editor displays an ! (exclamation point).

3. Exiting the ed Editor

q	The q (quit) subcommand exits the ed editor after checking whether the buffer has been saved to a file after the last changes were entered. If the buffer has not been saved to a file, the q subcommand displays the ? (question mark) message. Enter the q subcommand again to exit the ed editor anyway. The changes to the current file are lost.
Q	The Q (Quit) subcommand exits the ed editor without checking whether any changes were made since the buffer was saved to a file. Any changes made to the buffer since the last save are lost.

(a) To Quit after Checking for Edits

1. Type the following subcommand :

    q

2. If the ed editor displays a ?, type one of the following subcommands :

    To save changes before quitting, type :

    w

    then press the Enter key.

    To quit without saving changes, type :

    q

3. Press the Enter key.

(b) To Quit and Discard Edits

1. Type the following subcommand :

    Q

2. Press the Enter key. Any changes made to the buffer since the last save are lost.

4. Requesting Help

h	The h (help) subcommand provides a brief help message for the most recent ? diagnostic or error message displayed.
H	The H (Help) subcommand causes the ed editor to display help messages for all subsequent ? diagnostic messages. The H subcommand also explains the previous ? if one existed. The H subcommand alternately turns this mode on and off; it is initially off.

(a) To Start or Stop Displaying Help Messages

Enter the following subcommand :

    H

The help messages are displayed or not displayed for ? responses from the ed editor, depending on the previous setting.

(b) To Display the Last Help Message

Enter the following subcommand :

    h

A help message is displayed for the last ? response from the ed editor.

(c) Character Class Support in the ed Editor

In standard Patterns expression, a range expression matches the set of all characters that fall between two characters in the collation sequence of the current locale. The syntax of the range expression is as follows :

```
[character-character]
```

The first character must be lower than or equal to the second character in the collation sequence. For example, [a-c] matches any of the characters a, b, or c in the En_US locale.

The range expression is commonly used to match a character class. For example, [0-9] is used to mean all digits, and [a-z A-Z] is used to mean all letters. This form may produce unexpected results when ranges are interpreted according to the collating sequence in the current locale.

Instead of the preceding form, use a character class expression within [ ] (brackets) to match characters. The system interprets this type of expression according to the character class definition in the current locale. However, you cannot use character class expressions in range expressions.

The syntax of a character class expression is as follows :

```
[:CharacterClass:]
```

That is, a left bracket, a colon, the name of the character class, another colon, and then a right bracket.

The following character classes are supported in all locales :

`[:upper:]`	`Uppercase letters`
`[:lower:]`	`Lowercase letters`
`[:alpha:]`	`Uppercase and lowercase letters`
`[:digit:]`	`Digits`
`[:alnum:]`	`Alphanumeric characters`
`[:xdigit:]`	`Hexadecimal digits`
`[:punct:]`	`Punctuation character (neither a control character nor alphanumeric)`
`[:space:]`	`Space, tab, carriage return, new-line, vertical tab, or form feed character`
`[:print:]`	`Printable characters, including space`
[:graph:]	Printable characters, not including space
[:cntrl:]	Control characters
[:blank:]	Space and tab characters

The brackets are part of the character class definition. To match any uppercase ASCII letter or ASCII digit, use the following regular expression :

```
[[:upper:] [:digit:]]
```

Do not use the expression [A-Z0-9].

A locale may support additional character classes.

The newline character is part of the [:space:] character class but will not be matched by this character class. The newline character may only be matched by the special search characters $ (dollar sign) and ^ (caret).

Ed return the following exit values :

0 `Successful completion.`

>0 `An error occurred.`

3.21 EDITING WITH ED

The first two characters of the most common activity with a computer for programmers. Programmers are all the time editing source code. And they need a powerful editor to do this job. Hence the flame wars between the emacs and vi factions. But we have nothing to do with either. We are going to be talking about ed, the powerful line editor.

For a line editor, the terminal settings do not matter. Moreover in constrained embedded environments or when you boot single user in OpenBSD, you may love ed.

Let us cut to the chase now.

```
$ ed /etc/fstab
```

opens the file /etc/fstab in ed for reading and writing.

Once you are inside ed, you can view the file under edit with the keystrokes:

```
1,$n
```

with line numbers or

```
1,$p
```

without the line numbers.

Good. You can read the file now. How to write to it?

You do that with a funny sequence of commands and input text.

To start writing you have to go to the line which you want to change, add or delete.

This is easily done by simply typing the line number in UNIX spirit. Say little, but say enough.

So assume that you have a file of 1000 lines, saying

```
444
```

so ed will know that you want to be at line 444.

You can inspect the line number with the command.

```
.=
```

Now you wish to add a line at 444.

```
a
```

The new line added

```
.
```

will complete the action.

Now you have a line "The new line added" at line number 445.

You can verify the veracity of my statement by typing

```
1,$n
```

The other thing to do is when you wish to delete a line or range of lines.

You can do that easily with ed.

```
1,10d
```

will delete lines 1 to 10.

Remember, when we added a line, we prefixed the activity with a command :

```
a
```

and we suffixed with a

```
.
```

on a line all by itself. This is similar to the command line mail program which knows that you stopped typing the mail when you typed

```
.
```

on a line by itself.

You can also move and copy lines easily with ed.

```
1,5m$
```

will move the lines 1 to 5 to the end of the text file.

You can copy with the t operator.

```
1,5t20
```

will copy the lines 1 to 5 to line 20.

You can read shell commands into the text file with

```
r !ls
```

and you can also read and write to multiple files with

```
r /tmp/readfile.txt
```

and

```
w /tmp/newfile.txt
```

Once you are done with editing, kindly save and exit

```
wq
```

or quit without saving with just

```
q
```

on a line.

ed is a very powerful editor as it does all its actions in an internal buffer in memory. All your changes remain in main memory and they don't get written to the file until you explicitly write using a

```
w
```

command.

Now you may think that one of my opening statements on ed being independent of vi or emacs is actually false.

Yes, ed is related to vi.

The

```
wq
```

is very similar to the

```
:wq
```

command in vi or vim.

And you can also search for patterns using

```
/pat
```

operator in ed.

ed is very powerful and this article should get you started with this cryptic but very handy and powerful editing tool.

3.22. FEATURES OF ED

Features of ed include :

(i) Available on nearly all computer systems

(ii) A modal editor supporting command mode, text mode and viewing mode

(iii) Support for regular expressions

(iv) Powerful automation can be achieved by feeding commands from standard input

☞ SUMMARY

A good screen editor would have the simplicity and features of UNIX system line editor, ed. It would support its use of regular expression and its use of regular expression and its sophisticated search and substitute capabilities. vi is flexible, high-powered editor; They provide for sophisticated entering, modifying, and deleting of text.

The features of vi that have been highlighted so far are good enough for a beginner who should not proceed any further before mastering most of them. There are many more functions that make vi a very powerful editor. vi is a superset of ed, and it contains all of ed's features and syntax. In addition, vi provides extensions of its own that provide for customizing the editor. Users who are familiar with ed will find it easy to use vi because there are many basic similarities.

☞ EXERCISE

1. Discuss different operaional modes of vi editor. Write vi commands to perform the following :
   (i) Generate line numbers for the file being edited currently
   (ii) Delete three lines downward from current line.
   (iii) Mark a block of lines and perform cut and paste
   (iv) Save the currently opened file with a new name
   (v) Replace a word with new text. (UPTU 2005)
2. What is vi?
3. What is difference between command mode and insert mode?
4. How do I quit without saving to?
5. How do I insert a file in vi?
6. Why is vi an important UNIX text editor?
7. Add a new user to your test file (**Hint :** YYP)

❒❒❒

CHAPTER 4

GENERAL PURPOSE UTILITIES

4.1 INTRODUCTION

An instruction or direction to a computer or device to perform a specific task. This instruction is called command. Commands come in different forms. They can be:

- Special words that a program understands.
- Function keys
- Choices in a menu
- Buttons or other graphical objects on your screen

Every program or task that interacts with people responds to a specific set of commands. The set of commands and the syntax for entering them is called the user interface and varies from one program to another.

In UNIX there are basically only two concepts—files and directories. Most things we would consider to be something other than these two concepts in a Windows environment are probably files in UNIX. This actually makes for a very intuitive environment where if we can do something to a file then we can do it to just about anything.

By using UNIX commands, we can create and manage files containing a variety of data. These commands allow us to move around the system's file tree to create, copy, rename, move, and delete etc. The commands that we will use in this section are listed below and must be used in lower case. Because UNIX is case sensitive. In fact all commands are in lower case, and no commands have ever been created which use upper case characters.

To make our request understandable to the UNIX system. We must present each command in command line syntax. This syntax defines the order in which you enter the various components of a command line.

The command line syntax

$command [option(s)] [argument(s)]

Where command is the name of the command we wish to run, options modified how the command works and arguments specified the data on which the command is to be operate.

Rules of commands

You should follow several rules with UNIX commands:

1. UNIX commands are case-sensitive, but maximum commands in lowercase.
2. UNIX commands can only be entered at the shell prompt.
3. UNIX command lines must end with a ENTER.
4. UNIX options often begin with a "-" (minus sign).

More than one option can be included with many commands.

4.2 INTERNAL AND EXTERNAL COMMANDS

Some commands that we type are internal, built into the shell. For example, the cd command is built-in. That is, the shell interprets that command and changes your current directory for us. The ls command is an external program stored in the file /bin/ls.

The shell does not start a separate process to run internal commands. External commands require the shell to fork and exec a new subprocess. This takes some time, especially on a busy system.

When we type the name of a command, the shell first checks to see if it is a built-in command and then executes it. If the command name is an absolute pathname beginning with root, like /bin/l. There is no problem, the command is likewise executed. If the command is neither built-in, nor specified with an absolute pathname, the shell looks in its search path for an executable program or script with the given name.

The search path is exactly what its name implies: a list of directories that the shell should look through for a command whose name matches what is typed.

The search path is not built into the shell. It's something we specify in our shell setup files.

UNIX system programs are kept in directories called /bin and /usr/bin, with additional programs usually used only by system administrators in /etc and /usr/etc. Users often also have their own directories where the commands and scripts are kept, such as /usr/local/bin.

The search path is stored in an environment variable called PATH. A typical PATH setting might look something like this:

```
PATH=/bin:/usr/bin:/usr/bin/X11:/usr/ucb:/home/tim/bin:
```

The path is searched in order, if there are two commands with the same name, the one that is found first in the path will be executed.

You can add new directories to search your path on the fly, but the path is usually set in shell setup files.

We can discuss the unix common commands in a single chapter. These are given below.

4.3 GENERAL PURPOSE UTILITIES

1. cal : Display the calendar.

2. date : Display system date.
3. echo : Echo argument.
4. script : make typescript of terminal session
5. mailx : mail interactive message processing
6. passwd : Change your password
7. who : Know the all login users
8. who am I : Know the itself.
9. uname : Know the machine's name./ Find the name of the Unix.
10. tty : Know your terminal
11. stty : Set the terminal characteristics.
12. cat : Create file, Display content of file, Concatenate files.
13. cp : copy files
14. rm : Delete file or remove filess
15. mv : Renaming files.
16. more : view of file or paging output/ read file.
17. file : determine file type.
18. wc : Counting line, Word and character.
19. ls : Listing Files.
20. od : dump file in octal and other formats
21. cmp : Comparing two files.
22. comm : What is common
23. diff : Conversing one file to another.
24. lpr : standard print command or sends requests to line printer.
25. banner : Graphic size
26. man : Manual command for help
27. pwd : Display current directory.
28. mkdir : Make Directories.
29. cd : Change Directories.
30. rmdir : Remove Directories
31. chmod : Changing file permissions
32. touch : Changing the file stamps
33. umask : Default file permissions
34. grep : search for a string or word in a file.
35. pr : Paginating files.
36. head : Displaying the beginning of a file.
37. tail : Displaying the End of a file.
38. cut : Cutting a file Vertically.
39. paste : Pasting files.
40. sort : Sorting a file.
41. tr : Translating characters

42. find : Locating files.
43. dircmp : Directory Comparison
44. tee : For save the output of the command in a file
45. spell : Check spelling
46. dos 2 UNIX : UNIX text file format convertor
47. UNIX 2 dos : dos text file format convertor
48. grip : compress or expand files
49. gunzip : compress or expand files
50. zip : pakage and compress files
51. unzip : Extract compressed files
52. exit : Exit status of command.
53. finger : User information lookup program.
54. tput and clear : Clear the screen.
55. expr : Convert character string into numerical
56. lock : Lock your system.
57. write command.
58. wall command.
59. mesg command.
60. talk command.
61. bc : Calculator.
62. factor : Factorise the number.
63. ps : Process status.
64. nohup : Log out safely.
65. kill : Terminate the process.
66. crontab : Running jobs periodically.
67. ftp : connect to a remote machine to download or upload files
68. ncftp : especially good for downloading files via anonymous ftp.
69. rsh : remote shell
70. setenv : set an environment variable
71. tar : create an archive, add or extract files
72. env : Display all variables
73. CPIO : Backup file
74. telnet : log in to another machine.
75. Ln : Command
76. pipe : create pipe

GENERAL PURPOSE UTILITIES

1. Cal : The Calendar (Display the Calendar)

The cal command displays a calendar of the specified month or a complete year.

```
Syntax
```

```
$cal [month][year]
```

The year parameter names the year for which you want a calendar. The cal command can display a calendar for any year from 1 through 9999; we must enter the full year rather than just the last two digits. The month parameter identifies the month for which you want the calendar. It can be a number from 1 to 12 (indicating January to December). If we specify neither the Year nor the Month parameter then the cal command displays the current month. If we specify only one parameter then the cal command assumes the parameter is the year parameter and displays the calendar for the indicated year. Home

For example:

```
$cal
```

The output shows the current date as well as the calendar of preceding, current and succeeding month. Display complete calendar of the current year.

$cal 12 2007

December 2007

M	T	W	Th	F	S	S
					1	2
3	4	5	6	7	8	9
10	11	12	13	14	15	16
17	18	19	20	21	22	23
24	25	26	27	28	29	30
31						

The 12 here indicates December. In place of 12 we could have used dec.

$ cal 2007

Display the complete calendar of year 2007.

2. date : Display Both Date and Time (Display System Date)

In UNIX date command display the system date as well as time.

Syntax

```
$date [Enter]
```

For example, type the four characters in date and press Enter key.

```
$date
Mon Dec 3 12:28:34 IST 2007
$
```

Date is the valid command in UNIX and it displays both date and time. The date command instructs the machine to display the current date time. Most UNIX commands are represented as files in the system.

3. echo : Echo Argument

The echo command echoes its arguments.

Here are some examples:

```
$echo this
this
```

```
$echo $EDITOR
/usr/local/bin/mca
$echo $PRINTER
us101-dotmatrix
```

Things like PRINTER are so-called environment variables. This one stores the name of the default printer- the one that print jobs will go to unless you take some action to change things. The dollar sign before an environment variable is needed to get the value in the variable. Try the following to verify this:

```
$echo PRINTER
PRINTER
```

4. script : make typescript of terminal session

```
script - make typescript of terminal session
```

Syntax

```
$script [option] [file]
```

Script makes a typescript of everything printed on your terminal. It is useful for students who need a hardcopy record of an interactive session as proof of an assignment, as the typescript file can be printed out later with lpr command.

If the argument file is given, script saves all dialogue in file. If no file name is given, the typescript is saved in the file typescript.

Options

-a

Append the output to file or typescript retaining the prior contents.

-f

Flush output after each writes.

This is nice for telecooperation: One person does `mkfifo tanu; script -f tanu' and another can supervise real-time what is being done using `cat tanu'.

-q

Be quiet.

-t

Output timing data to standard error. This data contains two fields, separated by a space. The first field indicates how much time elapsed since the previous output. The second field indicates how many characters was output this time. This information can be used to replay typescripts with realistic typing and output delays.

5. mailx : mail interactive message processing

mailx helps you read and send electronic mail messages. It has no built-in facilities for sending messages to other systems, but combined with other programs , it can send messages to other systems.

The command line

```
$mailx [options] user user user ...
```

sends a mail message to the given users. If you do not specify any users on the command line, mailx lets you read incoming mail interactively.

Options

(i) mailx accepts the following options when you are reading messages

-e

checks to see if you have any messages waiting to be read. With this option, nothing is displayed. If you have waiting messages, mailx exits with a successful status return; otherwise, mailx exits with a failure return.

-f

looks for messages in the file given by the optional filename on the command line instead of in your system mailbox. If you do not specify filename, mailx reads messages from the file specified by the MBOX variable.

-H

displays only the header summary of a message.

-N

does not display the header summary of messages.

-u user

looks for messages in the system mailbox of the specified user. This only works if you have read permission on the user's system mailbox.

(ii) mailx accepts the following options only when you are sending messages

-F

records your message in a file with the same name as the first user specified on the command line. This option overrides the record variable, if it has been set.

-h number

indicates how many hops a message has already made from one machine to another . This option is not intended for most users; some network mail software uses the option to prevent infinite loops

-p priority

sets the priority of the mail message to be sent. priority can be low, normal, or high.

-r address

passes the given address to network mail software. If this option is present, it disables all input mode commands.

-s subject

uses the given subject string in the Subject heading line of the message. If the subject contains spaces or <tab> characters, the string should be enclosed in double quotes (" ") or apostrophes (' '). If you specify this option on the command line, mailx does not prompt you to enter a Subject line when you type in the text of the message.

-U

converts the address from UUCP style to internet standards. This overrides the effect of the conv environment variable.

mailx accepts the following options when you are sending or reading messages :

-i

ignores interrupts (for example, pressing BREAK or CTRL-C).

-m

uses Microsoft Exchange compatibility mode. In this mode, the From line is :

From : username

and the Date line is

Date : DayofWeek, DayofMonth Mon Year HH:MM:SS timezoneoffset

-n

does not initialize your mailx session from the system's ROOTDIR/etc/mailx.rc file. For more information about this file.

You can use mailx in many ways, and these are discussed later; however, you must first understand mailx's normal pattern of behavior.

The simplest command to send a message is

```
$mailx address address address ...
```

where each address names someone who is to receive the message. The simplest kind of address is the login name of someone else who uses your machine.

You can also send messages as input to commands. To do this, use an address that consists of an or-bar (|) followed by a command line that invokes the appropriate command. Enclose this whole address in apostrophes. For example,

```
$mailx robin '|cat >save'
```

mails a message to robin and also copies the message into a file called save.

After you type in the command to send a message, mailx asks you to enter the Subject of the message (this brief description of what the message is about can be up to 256 characters long). You can now type in the text of your message. Your message can consist of any number of lines, and may include blank lines. When you finish entering the message, type a line consisting only of a tilde (~), followed by a dot (.); then press ENTER. This tells mailx that the message is ready to be sent.

Suppose you write a script that sends a file using mailx.

```
$ mailx username@yahoo.com
```

If you can use ksh, $(expr) is better than `expr`.

Expressions like this can only be evaluated by the shell, not within applications.

```
$mailx -s $(date)
```

etc should work.

You don't have mailx to test right now, but you could try something like

```
$mailx -s $(date) < tanmayfile.txt
```

A message consists of a sequence of header lines followed by the body of the message. The header lines tell who sent the message, the time and date that it was sent, its subject, and so on. mailx automatically creates header lines.

Some of the common header lines are :

Cc : name name ...

Stands for carbon copies. This indicates that copies of this message are to be sent to the specified recipients. The names of these recipients appear in the header lines of everyone receiving the message.

Bcc: name name ...

Stands for blind carbon copies. This is similar to Cc: but the names of people receiving carbon copies do not appear in the header lines of the message. Recipients do not know that these people received a copy of the message.

Subject : text

Gives the subject of the message.

To : name name ...

Gives the names of people who were sent the message directly.

6. passwd: Change your password

Syntax

```
$passwd[Enter]
```

You can change your password at any time. You should change it the first time that you log in, and we recommend that you change it on a regular basis. At the command prompt, type passwd.You will be prompted to enter your old password and be asked twice to enter your new password. Neither your old nor new password will appear on the screen as you type.

For example, changing your password from arya80 to dk80 will look similar to the following example, except that the keystrokes for you old and new password will not be echoed on the screen.

```
passwd
current password: arya80
New password (? for help): dk80
New password (again): dk80
Password changed for userid
```

7. who : Who are the users (Know the all login users)

Syntax

```
$who [Enter]
```

UNIX is a multi-user operating system, so multiple users can use it concurrently. We might be interested in knowing the people who are using the system. Use the who command:

```
$who
Arya    pts/1    Dec    3    09:30
Dev     pts/2    Dec    3    09:31
Anu     pts/3    Dec    3    09:32
$
```

There are currently three users- Arya, Dev and Anu. These are actually the usernames or user-ids they are used to log in. The second field shows the name of the terminal the user is working on. The third field shows the date and time of the log in.

8. who am I : Print current userid (Know the itself)

Syntax

```
$who am I [Enter]
```

If we want to know the name of itself then write the "whoami" command on the dollar prompt ($) and just press the Enter key.

For example

```
$who am i
Arya        pts/1        Dec        3        09:30
```

9. uname : Know the machine's name/Find the name of the Unix

Syntax

```
$uname[Enter]
```

If we have to find the name of the UNIX system then we can use the command uname.

For example

```
$uname
Linux
$
```

10. tty : The Terminal Name

To display the name of your terminal, use the tty command.

For example, at the prompt, type the following:

```
$tty
```

The system displays information similar to the following:

```
/dev/tty01
```

In this example, tty01 is the name of the terminal, and /dev/tty06 is the device file that contains the interface to this terminal.

In tty command we can also use some option and then display terminal.

Syntax

```
$tty options
```

Options

- -l will print the synchronous line number.
- -s will return only the codes: 0 (a terminal), 1 (not a terminal), 2 (invalid options)

11. stty : Set the Terminal Characteristics

stty is a command for setting and reporting terminal I/O options.

Syntax

```
$stty [ -a ] [ -g ] [ Options ]
```

The stty command sets certain I/O options for the device i.e. the current standard input. This command writes output to the device i.e. the current standard output.

When we redirect standard input from a tty device by entering

```
$stty -a </dev/ttyx
```

where a and g is:

-a :- Writes the current state of all option setting to standard output.

-g :- Writes option setting to standard output in a form usable by another sty command.

For example:

Prints information about your terminal settings. The information printed is the same as if you had entered stty while interacting with a shell.

```
$stty -a
```

Prints information about all of your terminal settings.

12. The cat Command: Create File, Display content of File, Concatenate Files

You can create a file without a text editor by using the cat command (short for concatenate) and the ">" (redirect output) symbol. To create a file using the cat command, type:

```
$cat > filename
```

where filename is the name you wish to give the file. The command cat generally reads in a file and displays it to standard output. When there is no filename directly following the command, cat treats standard input as a file. The ">" symbol will redirect the output from cat into the new filename you specify.

This is one of the most flexible Unix commands. We can use to create, view and concatenate files. The redirection symbol is the greater than sign '>'. If the file does not exist then it is created for you.

For first example we create a file called "deep"

```
$cat >deep
red
green
blue
<control-D>
$
```

<control-D> stands for "hold the control key down, then tap 'd'". The symbol > tells the computer that what is typed is to be put into the file deep.

To view a file we use cat in a different way:

```
$cat deep
red
green
blue
$
```

Information however can be appended to the end of a file by using two greater-than signs '>>', but the file must exist for this to work. If we wish to add text to an existing file we do this:

```
$ cat >>deep
white
black
<control-D>
$
```

Now suppose that we have another file tmp that looks like this:

```
$cat tmp
cat
dog
$
```

The cat command is commonly used to print the contents of files. Its name arises from the ability to take many file name arguments and concatenate them (combine files). Thus to merge several files one after the other into a single file: Then we can join deep and tmp like this:

```
$cat deep tmp >deep2
```

We could check the number of lines in the new file like this:

```
$ wc -l dict2
     7
```

The command wc counts things - the number of characters, words, and line in a file.

13. The cp Command: Copy Files

Use cp to copy files or directories.

Syntax:

```
$cp [Source file] [Target file]
```

For example Copy the file1 into file2

```
$cp file1 file2
```

This makes a copy of the file file1.

```
$cp ~/poems/dk .
```

This copies the file dk in the directory poems to the current directory. The symbol "." stands for the current directory. The symbol "~" stands for the home directory.

14. The rm Command: Delete File or Remove Files

Use rm command to remove files from your directory.

Syntax:

```
$rm [file name]
```

For example, to remove the file file1

```
$rm file1
     remove file1? y
$rm letter*
     remove letter1? y
     remove letter2? y
     remove letter3? n
$
```

The first command removed a single file. The second command was intended to remove all files beginning with the string "letter." However, user decided not to remove letter3.

Options

- rm file1 will delete file named file1.

- rm -f option will remove write-protected files without prompting.
- rm -r option will delete the entire directory as well as all the subdirectories, very dangerous command.

15. The mv Command : Renaming Files or Moving Files

Use this command to change the name of file and directories.

Syntax

```
$mv [old file] [new file]
```

For example, to move the data file1 to file2

```
$mv file1 file2
```

The file that was named file1 is now named file2

16. The more Command: View of File or Paging Output/ Read File

More is a command used to read text files.

Syntax

```
$more [file name]
```

For example, we could do this:

```
$more poems
```

The effect of this to let you read the file "poems". It probably will not fit in one screen, so you need to know how to "turn pages". Here are the basic commands:

- **q ---** quit more
- **spacebar ---** read next page
- **return key ---** read next line

For still more information, use the command man more.

17. file :

File determine file type each argument in an attempt to classify it. There are three sets of tests performed in this order :

⇒ files system test

⇒ magic number tests and

⇒ language tests

The first test that saceeds causes the file type to be printed

Syntex

$ file [– + anme file] [– m magic file] file.

The file type printed will usually contain one of the words, text, executeable or data.

For example,

$ file file.c

file.c : C program text.

18. The wc Command: Counting Line, Word and Character

Use this command to count the number of characters, words, and lines in a file.

Syntax:

```
$wc [option] [file name]
```

For example, that we have a file deep with contents

```
$cat deep
red rojo
green verde
blue azul
white blanco
black negro
Then we can do this
$wc deep
5 10 56 tmp
```

This shows that deep has 5 lines, 10 words, and 56 characters.

The word count command has several options, as illustrated below:

```
$wc -l deep
5 tmp
$wc -w deep
10 tmp
$wc -c deep
 56 tmp
```

The wc command counts the characters, words or lines in a file depending upon the option.

Options

- wc -l filename : will print total number of lines in a file.
- wc -w filename : will print total number of words in a file.
- wc -c filename : will print total number of characters in a file.

19. The ls Command: Listing Files

Use ls to see what files you have. Your files are kept in something called a directory.

```
$ls
     foo       letter2
     foobar    letter3
     letter1   maple-assignment1
$
```

Note that you have six files. There are some useful variants of the ls command:

```
$ls l*
     letter1 letter2 letter3
$
```

Note what happened: all the files whose name begins with "l" are listed. The asterisk (*) is the "wildcard" character. It matches any string.

Options

- **ls :** will list all the files in your home directory, this command has many options.
- **ls -l :** will list all the file names, permissions, group, etc. in long format.
- **ls -a :** will list all the files including hidden files that start with . .
- **ls -lt :** will list all files names based on the time of creation, newer files bring first.
- **ls -Fx :** will list files and directory names will be followed by slash.
- **ls -R :** will lists all the files and files in the all the directories, recursively.
- **ls -R | more :** will list all the files and files in all the directories, one page at a time.

20. Od command : dump file in octal and other formats

Syntex

$ od [option] filename

octal dump displays contents as octal numbers. This can be useful when the output contains non-printable characters. For example, a filename may contain non-printable characters. It can also be handy to view binary files.

Options

-A, -address-radix=radix

Select the base in which file offsets are printed.

radix can be one of the following :

- **d** decimal
- **o** octal
- **x** hexadecimal
- **n** none (do not print offsets)

The default is octal. -j, -skip-bytes=bytes

Skip bytes input bytes before formatting and writing. If bytes begins with '0x' or '0X', it is interpreted in hexadecimal; otherwise, if it begins with '0', in octal; otherwise, in decimal. Appending 'b' multiplies it by 512, 'k' by 1024, and 'm' by 1048576. -N, -read-bytes=bytes

Only output up to bytes of each input file. Any prefixes and suffixes on bytes are interpreted as for the -j option. -t, -format=type

Select the format in which to output the file data. type is a string of one or more of the below type indicator characters. If you include more than one type indicator character in a single type string or use this option more than once, od writes one copy of each output line using each of the data types that you specified, in the order that you specified.

- named character
- **c** ASCII character or backslash escape
- **d** signed decimal
- **f** floating point
- **o** octal
- **u** unsigned decimal
- **x** hexadecimal

- **C** char
- **S** short
- **I** int
- **L** long
- For floating point (f) :
- **F** float
- **D** double
- **L** long double

-v, -output-duplicates

Output consecutive lines that are identical. By default, when two or more consecutive output lines would be equal, od outputs only the first line, and puts just an asterisk(*) on the following line to indicate that identical lines have been elided. -s, --strings[=bytes]

Instead of the normal output, output only string constants in the input, which are a run of at least bytes ASCII graphic (or formatting) characters, terminated by a NUL. If bytes is omitted, it defaults to 3.

-width[=bytes]

The number of input bytes to format per output line. It must be a multiple of the least common multiple of the sizes associated with the specified output types. If bytes is omitted, it defaults to 32. If this option is not given, it defaults to 16.

-help

Print a usage message and exit with a status code indicating success.

-version

Print version information on standard output then exit.

-a

Output as named characters. Equivalent to -t a.

-b

Output as octal bytes. Equivalent to -t oC.

-c

Output as ASCII characters or backslash escapes.
Equivalent to -t c.

-d

Output as unsigned decimal shorts. Equivalent to -t u2.

-f

Output as floats. Equivalent to -t fF.

-h

Output as hexadecimal shorts. Equivalent to -t x2.

-i

Output as decimal shorts. Equivalent to -t d2.

-l

Output as decimal longs. Equivalent to -t d4.

-o

Output as octal shorts. Equivalent to -t o2.

-x

Output as hexadecimal shorts. Equivalent to -t x2.

Example

```
$ dir | od -c | more
$ cat my_file | od -c |more
$ od my_file |more
```

Comparison of different outputs :

Show 16 first characters from a binary file (/bin/sh) as ASCII characters or backslash escapes (octal) :

```
$ od -N 16 -c /bin/sh
```

Output

```
0000000 177 E L F 001 001 001 \0 \0 \0 \0 \0 \0 \0 \0 \0
```

Show the same binary as named ASCII characters :

```
$ od -N 16 -a /bin/sh
```

Output

```
0000000 del E L F soh soh soh nul nul nul nul nul nul nul nul
nul
```

Show the same binary as short hexcadecimals :

```
$ od -N 16 -t x1 /bin/sh
```

Output :

```
0000000 7f 45 4c 46 01 01 01 00 00 00 00 00 00 00 00 00
```

Show the same binary as octal numbers :

```
$ od -N 16 /bin/sh
```

Output

```
$ 0000000 042577 043114 000401 000001 000000 000000 000000 000000
```

21. cmp : Comparing Two Files

Syntax:

```
$cmp [fine1] [file2]
```

For exmaple I have two different files oldfile and newfile

```
$cmp oldfile newfile
oldfile newfile differ: char 80, line 4
```

if we run cmp command on similar files nothing is returned.

-s command can be used to return exit codes. i.e. return 0 if files are identical, 1 if files are different, 2 if files are inaccessible.

This following command prints a message 'no changes' if files are same

$cmp -s fileone file1 && echo 'no changes' .

```
no changes
```

22. comm : What is Common

The comm command reads the File1 and File2 parameters and writes, by default, a three-column output to standard output.

Syntax

```
$comm [-1] [-2] [-3 ] file1 file2
```

-1	Suppress the output column of lines unique to file1.
-2	Suppress the output column of lines unique to file2.
-3	Suppress the output column of lines duplicated in file1 and file2.
file1	Name of the first file to compare.
file2	Name of the second file to compare.

The columns consist of:

- Lines that are only in File1
- Lines that are only in File2
- Lines that are in both File1 and File2.

Both File1 and File2 should be sorted according to the collating sequence specified by the current National Language environment.

For example:

```
$comm myfile1 myfile2
```

The above example would compare the two files myfile1 and myfile2.

23. diff : Conversing One File to Another

diff command will compare the two files and print out the differences between.

Here we have two ascii text files. fileone and filetwo.

Contents of fileone are

```
This is first file
this is second line
this is third line
this is different as;aryan
this is not different
```

contents of filetwo are

```
This is first file
this is second line
```

```
this is third line
this is different so many as;anu
this is not different
```

$diff fileone filetwo will give following output

```
4c4
< this is different as;aryan
---
> this is different so many as;anu
```

24. lpr : Standard Print Command or Sends Requests to Line Printer

The UNIX system includes a collection of several programs called lpr system, for printing documents and files. We can use it to print everything from simple text files.

Syntax

```
$lpr [file name]
```

By the using of lpr command we can print the any file.

For example:

```
$lpr nyfile.doc
request id is lsr1-143 (1 file)
```

The confirmation message from lpr returns a request id that we can use to check on the status of the job or to cancel it if we want.

We can print several files at once by including all of them in the argument to lpr.

For example:

```
$lpr anu*
```

Request id is lsr1-144 (4 files)

Prints all files whose name begins with anu.

25. The Banner Command

The banner command prints characters in a sort of ASCII art poster, for example to print ANU in big letters. We will type banner ANU at UNIX command line or in my script.

For example,

```
$banner ANU
```

```
     ###        ##         #    #       #
    #   #       # #        #    #       #
   #     #      #  #       #    #       #
  #       #     #   #      #    #       #
 #         #    #    #     #    #       #
 #### ####      #     #    #    #       #
#           #   #      #   #    #       #
#            #  #       ## #     #     #
#             # #        ##       #####
```

26. man : Manual Command for Help

The man command on UNIX provides access to online manual pages. We can use it to find out how to run a command in which we are interested, or sometimes to find out which command we need to run to perform a given operation.

Syntax

```
$man [command name]
```

Let we want to find out about ls, Use this command to read the ls manual page:

```
$man ls
```

Similarly, if we want to know about rm, the file-removal command, run the following command:

```
$man rm
```

To read about man itself, use this command:

```
$man man
```

27. pwd : (Print Working Directory) Display Current Directory

pwd command will print your home directory on screen, pwd means print working directory.

```
$pwd
/Home/mca01
  $cd homework
  $pwd
/Home/mca01/homework
  $ls
assign-1 assign-2 assign-3
  $cd ..
  $pwd
/Home/mca01
  $
```

Jan began by working in his "home" directory. Then his cd'd into his homework subdirectory. Cd means "change directory". He used pwd to check to make sure he was in the right place, and then used ls to see if all his homework files were there.. Then his cd' back to his home directory.

28. mkdir : Make Directories

Use this command to create a directory.

```
$mkdir tanmay
```

To get "into" this directory, do

```
$cd tanmay
```

To see what files are in tanmay, do this:

```
$ls
```

There shouldn't be any files there yet, since you just made it.

We can create files by using cat command.

29. The cd Command: Change Directories

Use cd to change directory. Use pwd to see what directory you are in.

```
$cd mydr
$pwd
$/u/me/all/mydr
$ls
novel poems
```

```
$cd novel
$pwd
$/u/me/all/mydr/novel
$ls
ch1 ch2 ch3 journal scrapbook
$cd ..
$pwd
$/u/me/all/mydr
$cd poems
$cd
$/u/me/all
```

All began in his home directory, and then went to his mydr subdirectory. He listed this directory using ls, found that it contained two entries, both of which happen to be diretories. He cd'd to the diretory novel, and found that he had gotten only as far as Chapter 3 in his writing. Then he used cd .. to jump back one level. If you wanted to jump back one level, then go to poems he could have said cd ../poems. Finally he used cd with no argument to jump back to his home directory.

30. The rmdir Command : Remove Directories

Use this command to remove a directory.

Syntax:

```
$rmdir [Directory Name]
```

For example, to remove a directory called "tanmay", do this:

```
$rmdir tanmay
```

A directory must be empty before it can be removed. To empty a directory, use rm.

Options:

- rm -r directory_name will remove all files even if directory is not empty.
- rmdir tanmay is how you use it to remove tanmay directory.
- rmdir -p will remove directories and any parent directories that are empty.
- rmdir -s will suppress standard error messages caused by -p.

31. (i) The chmod Command: Change the Mode or Permission

There are three different levels of permission:

1. **r** - read permission: allows users to read the contents of a file
2. **w** - write permission: allows users to change the contents of a file
3. **x** - execute permission: signifies a program or script file and allows users to execute (run) the file.

File permissions can be set for three different levels of ownership:

1. **u** - user - The user who created the file its its owner.
2. **g** - group - A group to which you are assigned, usually made up of those users engaged in similar activities, and who need to share files among themselves.
3. **o** - others - All other users on the system.

Thus you can set read, write, and execute permissions for the three classes of ownership:

```
-rxw rwx rwx
 |     \    \
 |      \    \
 |       \    \
owner   group   other
```

- **Permissions:** The characters indicate whether you allow users to read, write and execute a file.
- **Links:** The number of links a file has indicates how many different files actually point directly to the file you are looking at. A link count of 1 is normal, whereas a count of say 2 would indicate that two filenames point to the same actual file data - more of this later.
- **Owner:** Who is the file owned by
- **Group:** Which group is the owner of the file in? You will be in at least one group and may be in several.
- **Bytes:** Size of the file in bytes.
- **Creation date:** When was the file last written to?
- **File name:** What is the actual name of the file in question?

The values for any one digit are derived from adding the value of the permissions from the following list:

- **0** - no permission (-)
- **1** - execute permission only (x)
- **2** - write permission only (w)
- **4** - read permission only (r)

Notice that if you add all of the values from the values from the list, the result is 7. The number 7 represents read, write, and execute permission values. For read and write permissions only, the value would be (4+2 = 6). Read and execute permissions only would be (4 + 1 = 5).

This command is used to change the permissions of a file or directory.

For example to make a file myfile readable by everyone, we do this:

```
$chmod a+r myfile
```

or

```
$chmod 777 myfile
```

In this, we provide all the permissions to all means. owner, Group and others

```
$chmod 007 mynewfile
```

In this file, we provide all the permission to other and no permission to owner and group.

```
$ chmod 700 arya
```

We give the read, write and execution, permission to owner, but no permission to group and other.

Category can be omitted in which case the permissions specified are granted to all three categories such as user, group and other. If + is replaced by a - then those permissions are taken away from the categories.

As an example, consider this statement :

```
$chmod o+r tanmay
```

This grants other read permission to the file tanmay.

The command

```
$chmod +x tanmay
```

Grants everyone such as user, group and other execute permission, and the command

```
$chmod g+rwz tanmay
```

Gives category group read, write and execute permission.

Permissions change cumulatively in that the commands

```
$chmod +r fileanu
$chmod +w fileanu
$chmod +x fileanu
```

are equivalent to

```
$chmod +rwx fileanu
```

Remove permission from a file/directory

Following example removes read and write permission for the user.

```
$chmod u - rx filename
```

To make a file, e.g., a shell script mycommand executable, we do this

```
$chmod +x mycommand
```

Now we can run mycommand as a command.

To check the permissions of a file, use ls -l . For more information on chmod, use man chmod.

(ii) The Chgrp Command. : Change the Group

chgrp command is used to change the group of a file or directory.

You must own the file or be a superuser.

Syntax

```
$chgrp [options] newgroup files.
```

Newgroup is either a group Id or a group name located in /etc/group .

Options

- -h will change the group on symbolic links.
- -R recursively descend through directory changing group of all files and subdirectories.

(iii) The Chown Command : Change the Ownership

chown command to change ownership of a file or directory to one or more users.

Syntax

```
$chown options newowner files
```

Options

- -h will change the owner on symbolic links.
- -R will recursively descend through the directory, including subdirectories and symbolic links.

32. The touch Command: Changing the File Stamps

The touch command use for create the empty file. But it also allow us to change the modification and access time of the file.

Syntax

```
$touch option expression filename
```

When we used touch command without option or expression, both times are set to the by default current time.

The file is created if it does not exist, but not overwritten if it exist:

```
$touch myfile
```

When we used touch command with out options but with expression it changes both times (i.e. access time and modification time). If the touch command expression consists of an eight digit number using the format MMDDhhmm (Month, Day, Hour & Minute).

Now let we use the command on myfile:

$ls -l myfile

_rw_r_ _r_ _ Arya 1001 Dec 14 10:40 myfile
(Modification time)

$ls -lu myfile

_rw_r_ _r_ _ Arya 1001 Dec 14 12:40 myfile
(Access time)

In the given example first command shows the modification time, and the second command shows the access time of the file. But the touch command can change the both modification and access times:

$touch 12161245 myfile;ls -l myfile

_rw_r_ _r_ _ Arya 1001 Dec 16 12:45 myfile

$ls -lu myfile

_rw_r_ _r_ _ Arya 1001 Dec 16 12:45 myfile

We can also change two times individualy with -m option, we can alter the modification time alone. The -a option changes the access time.

```
$touch -a myfile
```

33. The umask Command : Default File or Directory Permissions

When we create directories and files, the permission assign to them depend on the system's default setting. The UNIX system has the following default setting for all directories and files:

- rwxrwxrwx (777) for all directories
- rw-rw-rw- (666) for all regular files

We don't see this permission when we create a directory or file. This default is transformed by subtracting the user mask from it to remove one or more permissions.

We can evaluate the current value of the mask by using umask command without argument.

```
$umask
022
```

This is an octal number which is to be subtracted from the system default to obtain the actual default. This becomes 755(777 - 022) for directory and 644 (666 - 022) for ordinary files. When we create a file on the system, it will be having the permission rw-r-r--.

34. The grep Command: Search for a String or Word in a File

Use this command to search for information in a file or files.

Syntax:

```
$grep option filename
```

For example, suppose that we have a file deep whose contents are

```
red
green
blue
white
black
```

Then we can look up items in our file like this;

```
$grep red deep
     red
$grep white deep
     white
$grep color deep
     $
```

Notice that no output was returned by grep colour. This is because "colour" is not in our dictionary file.

Grep can also be combined with other commands.

For example, if one had a file of phone numbers named "ph", one entry per line, then the following command would give an alphabetical list of all persons whose name contains the string "Fred".

```
$grep Fred ph | sort
Alpha, Anu: 011-2600001
Beta, Arya: 0120-270001
Gamma, Ashu-George: 0120-28001
Zeta, Dev: 012-290001
```

The symbol "|" is called "pipe." It pipes the output of the grep command into the input of the sort command.

For more information on grep, consult

```
$man grep
```

Options

-C	Display only the number of lines in each file that contain a match
-i	Ignore the case
-l	Display file name only
-n	preceds each line by its line number in the file

35. The pr Command : Paginating Files

The pr command prepares a file for printing by adding suitable headers, footers and formatted text. We can use it with a file name as argument:

```
$pr myfile
Dec 14 10:35: 2007 myfile page1
```

Pr adds five lines of margin at the top and five at the bottom. The lower portion of the page has not been shown in the example for reason of economy. The header shows the date and time of last modification of the file along with the filename and page number.

Option

There are many option in pr command:

Option	Description
-l n	set the length of page to n lines
-wn	sets the width of a page to n characters
-h str	sets the header for every page to string str
-n	numbers lines in output
-o n	offset output by n spaces
-d	double space input
-t	eliminates footer, header and margin

For example

```
$pr -L 70 myfile        [page set to 70 lines]
$pr +L 12 myfile        [ start printing from page 12]
```

36. head : Displaying the Beginning of a File

The head command display the beginning line of the file. The head command by default will display the first 10 lines of a file.

Syntax

```
$head [option] [file name]
```

For example : If we want first 50 lines we can use

```
$head -50 filename
or for 37 lines
$head -37 filename.
```

37. tail : Displaying the End of a File

The tail command displays the end line of the file. The tail command by default will display the last 10 lines of a file.

Syntax

```
$tail [option] [file name]
```

For example:

If we want last 50 lines then we can use

```
$tail -50 filename.
```

38. cut : Cutting a File Vertically

The cut command selects a list of columns or fields from one or more files.

Option -c is for columns and -f for fields.

Syntax

```
$cut options [files]
```

for example if a file named testfile contains

```
this is firstline
this is secondline
this is thirdline
```

Examples:

```
$cut -c1,4 testfile, will print this to standard output
ts
ts
ts
```

It is printing columns 1 and 4 of this file which contains t and s (part of this).

Options

- -c list : cut the column positions identified in list.
- -f list : will cut the fields identified in list.
- -s could be used with -f to suppress lines without delimiters.

39. paste : Pasting Files

The paste command merges the lines of one or more files into vertical columns separated by a tab.

Syntax

```
$paste [file1] [file2]
```

For example if a file named testfile contains

```
this is firstline
```

and a file named testfile2 contains

```
this is testfile2
```

then running this command

$paste testfile testfile2 > outputfile

will put this into outputfile

```
this is firstline this is testfile2
```

it contains contents of both files in columns.

who | paste - - will list users in two columns.

Options

- -d'char' separate columns with char instead of a tab.
- -s merge subsequent lines from one file.

40. sort : Sorting a File

We can use the sort command to order the contents of files, sorting alphabetically, numerically, or by different fields. The sort command displays sorted file contents on a line-by-line basis. It compares the first characters in each line; if they are the same, it compares the next characters in turn, and so on through to the end of the line

Syntax

```
$sort [option] [filename]
```

For example if you have a file named testfile with these contents

```
zzz
aaa
1234
yuer
wer
qww
wwe
```

Then running

$sort testfile

will give us output of

```
1234
aaa
qww
wer
wwe
yuer
zzz
```

Options

- -b ignores leading spaces and tabs.
- -c checks whether files are already sorted.
- -d ignores punctuation.
- -i ignores non-printing characters.
- -n sorts in arithmetic order.
- -ofile put output in a file.
- +m[-m] skips n fields before sorting, and sort upto field position m.
- -r reverse the order of sort.
- -u identical lines in input file apear only one time in output.

41. tr : Translating Characters

The tr (translate characters) command is one of the true Unix filters. It copies its standard input to its standard output, while replacing the first character specified on the command line with the second character specified:

Syntax

```
$tr [ options ] [ set1 [ set2 ] ]
$ echo "hello world" | tr h k
kello world
```

Here are more full descriptions of those options:

1. **-c, --complement** Complement set1 with respect to the universe of characters whose ASCII codes are 01 through 0377 octal.
2. **-d, --delete** Delete characters in set1 but do not translate.
   tr --delete 'aeiou'
3. **-s, --squeeze-repeats** Replace sequences of the same character with one. Useful for deleting sequences of blanks.
4. **-t, --truncate-set1** Truncate set1 to the length of set2

42. find : Locating Files

Find command is an extremely useful command. we can search for any file anywhere using this command provided that file and directory you are searching has read write attributes set to you ,your, group or all. Find descends directory tree beginning at each pathname and finds the files that meet the specified conditions. Here are some examples.

Some Examples:

$find $HOME -print will lists all files in your home directory.

$find /work -name chapter1 -print will list all files named chapter1 in /work directory.

$find / -type d -name 'man*' -print will list all manpage directories.

$find / -size 0 -ok rm {} \; will remove all empty files on system.

conditions of find

- **-atime +n |-n| n** will find files that were last accessed more than n or less than -n days or n days.
- **-ctime +n or -n** will find that were changed +n -n or n days ago.
- **-depth** descend the directory structure, working on actual files first and then directories. You can use it with cpio command.
- **-exec commad {} \;** run the Unix command on each file matched by find. Very useful condition.
- **-print** print or list to standard output .
- **-name pattern** find the pattern.
- **-perm nnn**find files whole permission flags match octal number nnn.
- **-size n** find files that contain n blocks.
- **-type c** Find file whole type is c. C could be b or block, c Character special file, d directory, p fifo or named pipe, l symbolic link, or f plain file.

43. dircmp : Directory Comparison

The dircmp command compares two directories. If we have two directories in my home directory named dirone and dirtwo and each has 5-10 files in it. Then

```
$dircmp dirone dirtwo
Dec 14 10:06 2007 dirone only and dirtwo only Page 1
./cal.txt                  ./fourth.txt
./dohazaar.txt             ./rmt.txt
./four.txt                 ./te.txt
```

```
./junk.txt                ./third.txt
./test.txt
```

44. tee : Save out in a file

If we want to save the output of a command in a file, we would also like to see it on the terminal as it is being produced. A useful command which helps us achieve this is tee acts like a T-junction.

For example

```
$ls-l : tee fcounts
```

shows a full listing of the contents of the current directory on the terminal screen and also puts in the file fcounts.

45. spell : checking spelling

The spell command checks the words in a file against a directory file. It displays a list of words that it cannot either find in the directory or derive from one of the words in the directory.

The syntax of spell command is

```
$spell [filename]
```

For example, suppose we want to check the spelling of the file anu

```
$spell anu ↵
```

46. dos2unix : UNIX text file format converter

```
dos2unix - DOS/MAC to UNIX text file format converter
```

Syntax

```
$dos2unix[options] [-c convmode] [-o file ...] [-n infile outfile
...]
```

Options

The following options are available :

-h -help

Print online help.

-k -keepdate

Keep the date stamp of output file same as input file.

-q -quiet

Quiet mode. Suppress all warning and messages.

-V -version

Prints version information.

-c -convmode convmode

Sets conversion mode.

Examples

Get input from stdin and write output to stdout.

$dos2unix

Convert and replace anu.txt. Convert and replace tanu.txt.

$dos2unix anu.txt tanu.txt

$dos2unix -o anu.txt tanu.txt

Convert and replace anu.txt in ASCII conversion mode.

$dos2unix anu.txt -c iso tanu.txt

Convert and replace tanu.txt in ISO conversion mode.

$dos2unix -c ascii anu.txt -c iso tanu.txt

Convert arya.txt from Mac to UNIX ASCII format.

$dos2unix -c mac arya.txt tanu.txt

Convert and replace anu.txt while keeping original date stamp.

$dos2unix -k anu.txt

$dos2unix -k -o anu.txt

Convert anu.txt and write to dev.txt.

$dos2unix -n anu.txt dev.txt

Convert anu.txt and write to dev.txt, keep date stamp of dev.txt same as anu.txt.

$dos2unix -k -n anu.txt dev.txt

Convert and replace anu.txt. Convert tanu.txt and write to dev.txt.

$dos2unix anu.txt -n tanu.txt dev.txt

$dos2unix -o anu.txt -n tanu.txt dev.txt

Convert arya.txt and write to dev.txt. Convert and replace anu.txt. Convert and replace tanu.txt. Convert dk.txt and write to jp.txt.

$dos2unix -n arya.txt dev.txt -o anu.txt tanu.txt -n dk.txt jp.txt

47. unix2dos : dos text file format converter

unix2dos - UNIX to DOS text file format converter

Syntax

$unix2dos [options] [-c convmode] [-o file ...] [-n infile outfile ...]

Options

The following options are available :

-h -help

Print online help.

-k -keepdate

Keep the date stamp of output file same as input file.

-q -quiet

Quiet mode. Suppress all warning and messages.

-V -version

Prints version information.

-c -convmode convmode

Sets conversion mode.

-o -oldfile file ...

Old file mode. Convert the file and write output to it.

The program default to run in this mode. Wildcard names may be used.

-n -newfile infile outfile ...

New file mode. Convert the infile and write output to outfile.

File names must be given in pairs and wildcard names should NOT be used or you WILL lost your files.

Examples

Get input from stdin and write output to stdout.

$unix2dos

Convert and replace anu.txt. Convert and replace tanu.txt.

$unix2dos anu.txt tanu.txt

$unix2dos -o anu.txt tanu.txt

Convert and replace anu.txt in ASCII conversion mode. Convert and replace tanu.txt in ISO conversion mode.

$unix2dos anu.txt -c iso tanu.txt

$unix2dos -c ascii anu.txt -c iso tanu.txt

Convert and replace anu.txt while keeping original date stamp.

$unix2dos -k anu.txt

$unix2dos -k -o anu.txt

Convert anu.txt and write to dev.txt.

$unix2dos -n anu.txt dev.txt

Convert anu.txt and write to dev.txt, keep date stamp of e.txt same as anu.txt.

$unix2dos -k -n anu.txt dev.txt

Convert and replace anu.txt. Convert tanu.txt and write to dev.txt.

$unix2dos anu.txt -n tanu.txt dev.txt

$unix2dos -o anu.txt -n tanu.txt dev.txt

Convert arya.txt and write to dev.txt. Convert and replace anu.txt. Convert and replace tanu.txt. Convert dk.txt and write to jp.txt.

$unix2dos -n arya.txt dev.txt -o anu.txt tanu.txt -n dk.txt jp.txt

48. gzip command : compress or expand files

gzip reduces the size of the named files, each file is replaced by one with the extension **.gz,** while keeping the same ownership modes, access and modification times.

For example, the command

```
$gzip tanmay
```

Replace the file tanmay with the compressed file tanmay.gz. To decompress files that were encrypted using gzip, you can use gzip command with the -d(decrypt) option.

By default, gzip keeps the original file name and timestamp in the compressed file. These are used when decompressing the file with the **-N** option. This is useful when the compressed file name was truncated or when the time stamp was not preserved after a file transfer.

49. gunzip command : compress or expand files

You can also compressed by using gunzip command.

For example,

```
$gunzip tanmay
```

Or

```
$gunzip -d tanmay
```

Will replace the encrypted file tanmay.gz with the original file tanmay gunzip takes a list of files on its command line and replaces each file whose name ends with .gz, -gz, .z, -z, _z or .Z and which begins with the correct magic number with an uncompressed file without the original extension. gunzip also recognizes the special extensions .tgz and .taz as shorthands for .tar.gz and .tar.Z respectively. When compressing, gzip uses the .tgz extension if necessary instead of truncating a file with a .tar extension.

50. zip command : package and compress files

zip is a compression and file packaging utility for Unix. The zip command puts one or more compressed files into a single zip archive, along with information about the files (name, path, date, time of last modification, protection, and check information to verify file integrity). An entire directory structure can be packed into a zip archive with a single command. Compression ratios of 2:1 to 3:1 are common for text files. zip has one compression method and can also store files without compression. zip automatically chooses the better of the two for each file to be compressed.

When given the name of an existing zip archive, zip will replace identically named entries in the zip archive or add entries for new names.

For example, if tanmay.zip exists and contains tanu/file1 and tanu/file2, and the directory tanu contains the files tanu/file1 and tanu/file3, then :

```
$zip -r tanmay tanu
```

will replace tanu/file1 in tanmay.zip and add tanu/file3 to tanmay.zip. After this, tanmay.zip contains tanu/file1, tanu/file2, and tanu/file3, with tanu/file2 unchanged from before.

51. Unzip command : extract compress files

```
unzip - list, test and extract compressed files in a ZIP archive.
```

Syntax

```
$unzip [option] filename/directory name
```

unzip will list, test, or extract files from a ZIP archive, commonly found on MS-DOS systems. The default behavior is to extract into the current directory (and subdirectories below it) all files from the specified ZIP archive. A companion program, zip(1L), creates ZIP archives; both programs are compatible with archives created by PKWARE's PKZIP and PKUNZIP for MS-DOS, but in many cases the program options or default behaviors differ.

52. Exit : Exit Status of Command

The main use the exit command in the Unix terminate the present status.

Syntax:

```
$exit [Enter]
```

After this event simply terminate the command.

53. Finger : User Information Lookup Program

The finger command in the UNIX provide us with more information about other users on the system.

Syntax:

```
$finger [login name]
```

For example

```
$finger devanu
```

Will find out information about the user devanu, such as

```
Login name : devanu          Name : Anu Dev
Directory :/home/devanu      Shell :/usr/bin/bash
```

Last login Dec 07 10:15:10 on pts/01

Project : Currently I'm writing up my new book on UNIX.

The last line of the o/p from finger is the content of the file called .project. To create your own .project:

```
$cat >.project
```

My book all most finish and this book may be published in Jan 11.

```
Ctrl-d
```

54. tput : Term Information

The purpose of tput command Queries the terminfo database for terminal-dependent information.

Syntax

```
$tput [ -T Type ] [ CapabilityName {clear, init, longname,
reset} [ Parameters... ]]
```

For Using stdin to Process Multiple Capabilities

```
$tput [ -S ]
```

The tput command uses the terminfo database to make terminal-dependent information available to the shell. The tput command outputs a string if the attribute CapabilityName is of type string. The output string is an integer if the attribute is of type integer. If the attribute is of type Boolean, the tput command sets the exit value (0 for TRUE, 1 for FALSE), and produces no other output.

In addition to the capability names, the following strings are supported as arguments to the tput subroutine.

clear	Displays the clear screen sequence (this is also a capability name).
init	Displays the sequence that initializes the user's terminal in an implementation-dependent manner.

reset — Displays the sequence that will reset the user's terminal in an implementation-dependent manner.

longname — Displays the long name and the specified terminal (or current terminal if none specified).

-S — Uses stdin. This allow the tput to process multiple capabilities. When using the -S option, the capabilities cannot be entered on the command line. Enter ^D token finished.

-Ttype — Indicates the type of terminal. If -T is not specified, the TERM environment variable is used for the terminal.

Exit Status

This command returns the following exit values:

0 The requested string was written successfully.

1 Unspecified.

2 Usage error.

3 No information is available about the specified terminal type.

4 The specified operand is invalid.

>4 An error occurred.

Examples

1. To clear the screen for the current terminal, enter:

   $tput clear

2. To display the number of columns for the current terminals, enter:

   $tput cols

3. To display the number of columns for the aryaterm terminal, enter:

   $tput -Taryaterm cols

4. To initialize the current terminal, enter:

   $tput init

(ii) clear : Clear the screen.

Syntax:

```
$clear [Enter]
```

Clear command clears the screen and puts cursor at beginning of first line.

55. expr : Covert character string into numerical

Syntax

```
$expr expression
```

In UNIX shell programming Bourne shell has computing feature and this is expr command. This command combines two functions in one:

- Performs arithmetic operations on integers
- Manipulate string.

The expr command can perform the four basic operations as well as modulus functions:

For examples:

```
$a = 5 b = 2
$expr $a + $b
```

```
7
$expr $a - $b
3
$expr $a \* $b
10
$expr $a / $b
2
$expr 12 % 5
2
Length of the string
$expr "abcdefg"
7
```

56. lock : Lock the System

Syntax

```
$lock [-Timeout]
```

The lock command requests a password from the user, reads it, and requests the password a second time to verify it. In the entering, the command locks the terminal and does not relinquish it until the password is received the second time or one of the following occurs:

- The timeout interval is exceeded.
- The command is killed by a user with appropriate permission.

The timeout default value is 15 minutes, but this can be changed with the -Timeout flag.

-Timeout: Indicates the timeout interval in minute, as specified by the Timeout parameter. The default value is 14 minute.

Examples

1. To reserve a terminal under password control, enter:

   lock

   You are prompted for the password twice so the system can verify it. If the password is not repeated within 15 minutes, the command times out.

2. To reserve a terminal under password control, with a timeout interval of 10 minutes, enter:

   $lock -10

   /usr/bin/lock- Contain the lock command.

57. The Write Command

We can send a short message to the other a user with the write command.

Syntax

$write [login name]

Message......

....

Ctrl-d

For example

If your login name is anu then the command

$write arya

Hi arya, how are you?

Ctrl-d

Will display the following message on the arya's screen

Message from anu

Hi arya, how are you?

58. The Wall Command

Wall mean to say "write to all". This command can only be used by the super user.

Syntax:

$wall

Message format.......

....

Press CTRL-D

For example: If you want to display information to all connected users that tomorrow will be holiday then you do so using wall.

$wall

Dear friends tomorrow will be holiday.

CTRl-D

All users who are logged in, will hear a beep and see a message flashed on their screens:

Broadcast Message from root(tty/02) on UNIX Dec 12 10:35 2007

Dear friends tomorrow will be holiday.

59. The mesg Command

Both the command talk and write allow someone to type a message that will be displayed on your screen. You may be disturbed or you may find it disconcertrating to have message appear unexpectedly while you are working. The UNIX system provides the mesg command. This command allows you to accept or refuse message sent via talk and write

```
$mesg n
```

No permission for messages. Any one who tries to write a message you get the error message Permission denied.

```
$mesg y
```

Permission granted

60. The talk Command

A problem we have with write command is that our message can overlap each other. The talk command is the solution of this problem. We can communicate continue to each other.

Syntax

```
$talk [login name]
```

For example: If your login name is anu, and you type

```
$talk arya
```

The talk command notifies arya that you want to asks him/hir and speak with him/her to approve. Arya sees the following on his screen:

Message from talk_demo@abes at 10:15....

Talk: connection requested by anu@abes

Talk: respond with anu@abes

If arya respnd with talk anu@abes, talk splites on screen into upper and lower half.

You can send the reply write the message in the lower half. When you wish to end the session, press CTRL-D.

61. bc : Calculator

Calculator.

Syntax

```
$bc [-c] [-l] [file]
```

-c	Compile only.
-l	Define the math functions
file	Name of the file that contains the bc commands to be calculated this is not a necessary command.
/* and /*	Comment
sqrt (E)	Square root
quit	Exits the bc command.

Available Operators

+ - * / % ^

(+ Addition,- Subtraction,* Multiplication,/ Division, % is remainder;^ power)

++ -- (prefix and postfix; apply to names)

Examples

$bc cal.txt = Would result in 3, see below notes.

Within the cal.txt file you could have a simple statement such as:

/* Add the value 1+2 /*

1 + 2

quit

When running the above command we will receive the results of the cal.txt file. Which in this case would be 3.

62. factor : Factorise the number

Syntax

```
$factor [NUMBER]...
$factor OPTION
```

Print the prime factors of all specified integer Numbers.

For example:

```
$ factor 6
3 2
```

63. ps : Process Status

Syntax

```
$ps options
```

The ps command is probably the most useful command for systems administrators. It reports information on active processes.

options

- -a Lists all processes in system with all users except processes not attached to terminals.
- -e Lists all processes in system.
- -f Lists a full listing.
- -j print process group ID and session ID.
- -t List of all process from a particular terminal.

For examples:

If we want to see which process are running at any instant

$ps

PID	TTY	TIME	COMMAND
1234	tty/01	0.05	sh
1236	tty/01	0.00	ps

Unix assigns a unique number to every process. This number is called process ID or PID. The PIDs start with 0 and run up to a maximum of 32767.

```
Option -a :
$ps -a
Option -t
$ps -t tty/01
Option -f
$ps -f
Option -e
$ps -e
```

64. nohup : Log out Safely

The nohup command if added in front of any command will continue running the command or process even if you shut down your terminal or close your session to machine. Using this command we can submit the time consuming commands in the background, log out and leave the system and come next day to see our output ready.

For example,

```
$nohup sort empdetail >output.empdetail
27852
```

We can safely log out with out process getting terminated on logging out. After sorting our output would be ready in output.empdetail.

Another exmaple, if we want to run a job that takes lot of time and must be run from terminal and is called update_entries_tonight.

nohup update_entries_tonight will run the job even if terminal is shut down in middle of this job.

65. kill : Terminate the Process

We can terminate the currently running process by just hit the break or delete key, but we can not terminate the background process by hitting break or delete key. By using kill command we can terminate the background process.

Syntax:

```
$kill [PID]
```

For example, let we want to terminate process ID 2156

```
$kill 2156
2156 terminate
```

We can terminate the complete session by using the kill -9 command.

$ps

PID	TTY	TIME	COMD
1132	tty/01	0.05	sh
12345	tty/01	0.01	ps
12356	ty/02	0.25	sh

$kill -9 12356

A process group is a set of related processes. By using kill -0 command we can terminate all processes in the process group.

$kill -0 PID

66. crontab : Running Jobs Periodically

crontab command is used to schedule jobs. You must have permission to run this command by unix Administrator. Jobs are scheduled in five numbers, as follows.

Minutes	0-59
Hour	0-23
Day of month	1-31
month	1-12
Day of week	0-6 (0 is sunday)

For example you want to schedule a job which runs from script named backup_jobs in /usr/local/bin directory on sunday (day 0) at 11.25 (22:25) on 15th of month. The entry in crontab file will be. * represents all values.

25 22 15 * 0 /usr/local/bin/backup_jobs

The * here tells system to run this each month.

Syntax is

```
$crontab file
```

So, create a file with the scheduled jobs as above and then type crontab filename .This will scheduled the jobs.

67. ftp : connect to a remote machine to download or upload files

ftp command is used to execute ftp protocol using which files are transferred over two systems.

Syntax

```
$ftp options hostname
```

Options

- -d enable debugging.
- -g disable filename globbing.
- -i turn off interactive prompts.
- -v verbose on. show all responses from remote server.

ftp hostname by default will connect you to the system, you must have a login id to be able to transfer the files. Two types of files can be transferred, ASCII or Binary. bin at ftp> prompt will set the transfer to binary. Practice FTP by ftping to nic.funet.fi loggin in as anomymous with password being your e-mail address.

Use ftp to connect to a remote machine, then upload or download files. See also: ncftp

EXAMPLE 1

We'll connect to the machine fubar.net, then change directory to mystuff, then download the file homework11:

```
$ftp solitude
    Connected to fubar.net.
    220 fubar.net FTP server (Version wu-2.4(11) Mon Nov 5
    17:26:33 MDT 2007) ready.
    Name (solitude:carlson): tanmay
    331 Password required for tanmay.
    Password:
    230 User tanmay logged in.
ftp> cd mystuff
    250 CWD command successful.
ftp> get homework11
ftp> quit
```

EXAMPLE 2

We'll connect to the machine fubar.net, then change directory to mystuff, then upload the file collected-letters:

```
$ftp solitude
    Connected to fubar.net.
    220 fubar.net FTP server (Version wu-2.4(11) Mon 5 17:26:33
    MDT 2007) ready.
    Name (solitude:carlson): tanmay
```

```
    331 Password required for tanmay.
Password:
    230 User tanmay logged in.
$ftp> cd mystuff
    250 CWD command successful.
$ftp> put collected-letters
$ftp> quit
```

The ftp program sends files in ascii (text) format unless you specify binary mode:

```
$ftp> binary
$ftp> put arya
$ftp> ascii
$ftp> get bar
```

The file arya was transferred in binary mode, the file bar was transferred in ascii mode.

68. ncftp : especially Good for Downloading Files via Anonymous ftp.

Use ncftp for anonymous ftp that means you don't have to have a password.

```
$ncftp ftp.fubar.net
    Connected to ftp.fubar.net
$get jokes.txt
```

The file jokes.txt is downloaded from the machine ftp.fubar.net.

69. rsh : remote Shell

Use this command if you want to work on a computer different from the one you are currently working on. One reason to do this is that the remote machine might be faster. For example, the command

```
$rsh solitude
```

connects you to the machine solitude. This is one of our public workstations and is fairly fast. Just like telnet

70. setenv : set an Environment Variable

```
$echo $PRINTER
    labprinter
$setenv PRINTER myprinter
$echo $PRINTER
    myprinter
```

71. tar : Create an Archive, Add or Extract Files

tar command manupulate the individual character

The tar command creates an archive of files into a single file.

tar copies and restore files to a tape or any storage media.

Syntax

```
$tar [options] [file]
```

Examples:

$tar cvf /dev/rmt/0 /bin /usr/bin creates an archive of /bin and /usr/bin, and store on the tape in /dev/rmt0.

$tar tvf /dev/rmt0 will list the tape's content in a /dev/rmt0 drive.

$tar cvf - 'find . -print' > backup.tar will creates an archive of current directory and store it in file backup.tar.

Functions

- c creates a new tape.
- r append files to a tape.
- t print the names of files if they are stored on the tape.
- x extract files from tape.

Options:

- b n use blocking factor of n.
- l print error messages about links not found.
- L follow symbolic links.
- v print function letter (x for extraction or a for archive) and name of files.

Use create compressed archives of directories and files, and also to extract directories and files from an archive. Example:

```
$tar -tvzf anu.tar.gz
```

displays the file names in the compressed archive foo.tar.gz while

```
$tar -xvzf anu.tar.gz
```

extracts the files.

72. env : Display Environmental Variables

Displays your Unix environment variables.

Syntax:

```
$env
$env [-i] [variable=value...] [command argument...]
$env [-] [variable=value...] [command argument...]
```

If we call env with no arguments, it displays the environment that it received from its parent.

Arguments of the form

```
variable=value
```

let we add new variables or change the value in existing variables of the environment.

If we specify command, env calls command with the arguments that appear on the command line, passing the accumulated environment to this command. The command is executed directly as a program found in the search PATH and is not interpreted by a shell.

Options

-i

does not use the environment inherited by **env**

-

is an obsolete version of -i

Examples

Compare the output of the following two examples which illustrate the use of env:

```
$env foo=bar env
$env -i foo=bar env
```

Environment Variables

PATH

contains a list of directories to search when attempting to find command.

Options

Possible exit status values are:

0

Successful completion.

1

Failure due to any of the following

— insufficient memory

— name is too long

2

Invalid command line argument.

126

env found command but was unable to invoke it.

127

env was unable to find command.

Too many environment variables

The maximum number of environment variables that can be specified in a single env command is 512.

Examples

Typing env at the prompt would give something similar to.

HOME=/computerhope/public_html

PATH=/usr/local/bin:

LOGNAME=admin

HZ=100

TERM=vt100

TZ=MST7MDT

SHELL=/bin/csh

MAIL=/var/mail/computerhope

_INIT_UTS_PLATFORM=SUNW,SPARCstation-10

_INIT_UTS_RELEASE=5.7

_INIT_UTS_SYSNAME=SunOS

_INIT_UTS_VERSION=Generic_106541-08

EDITOR=pico -t

OPENWINHOME=/usr/openwin

MANPATH=/usr/man:/usr/local/man:/usr/openwin/man

LD_LIBRARY_PATH=/usr/local/lib:/usr/openwin/lib

PAGER=more

73. cpio : Back File System

The cpio command is useful to backup the file systems. It copy file archives in from or out to tape or disk, or to another location on the local machine.

Syntax

$cpio flags [options]

It has three flags, -i, -o, -p

$cpio -i [options] [patterns]

$cpio -i

- copy in files who names match selected patterns.
- If no pattern is used all files are copied in.
- It is used to write to a tape.

  $cpio -o
- Copy out a list of files whose names are given on standard output.

  $cpio -p
- copy files to another directory on the same system.

Options

- -a reset access times of input files.
- -A append files to an archive (must use with -o).
- -b swap bytes and half-words. Words are 4 bytes.
- -B block input or output using 5120 bytes per record.
- -c Read or write header information as Ascii character.
- -d create directories as needed.
- -l link files instead of copying.
- -o file direct output to a file.
- -r rename files interactively.
- -R ID reassign file ownership and group information to the user's login ID.
- -V print a dot for each file read or written.
- -s swap bytes.
- -S swap half bytes.
- -v print a list of filenames.

Examples

$ find . -name "*.old" -print | cpio -ocvB > /dev/rst8 will backup all *.old files to a tape in /dev/rst8

$ cpio -icdv "save"" < /dev/rst8 will restore all files whose name contain "save"

$ find . -depth -print | cpio -padm /mydir will move a directory tree.

74. telnet : log in to Another Machine

Use this command to log in to another machine from the machine you are currently working on. For example, to log in to the machine "solitude", do this:

```
$telnet solitude
```

75. Ln command

This command creates a hard link we can refer to the file either by its original filename or by the name given to it by the Ln Command.

The syntax of Ln command is

```
$Ln pathname
```

For example,

Example 1

```
$Ln /user/anu/anudir/arya
```

This command makes a link between the arya file in /usr/anu/anudir directory and the working directory. The file appears as arya in the working directory.

Example 2

```
$Ln /usr/anu/anudir/arya new arya
```

This command makes a link to the same file. In the above example, the file appears as newarya in the marking directory.

76. The Pipe Command : create pipe

You have seen how to use the > symbol to redirect output from standard output to a named file, and you have used the >> symbol to redirect output and append it to a named file without overwriting that file. However, by using what is termed a pipe, output from one command can be sent directly to another command as input. A pipe can link any two programs, provided the first program writes its output to stdout and the second program reads its input from stdin. The pipe symbol is represented as a (|). Its general syntax is as follows:

```
command1 | command2
```

command1 is the first command in the pipeline whose output is piped through the | symbol as input to command2. For example, suppose you want to know how many lines of your file contain when combined together. You could do the following:

```
$ cat file1 file2 > sample_file
$ wc -l < sample_file
121
```

With a pipeline, these two commands can be combined into a single line:

```
$ cat file1 file2 | wc -l
121
```

As we can see, using pipes eliminates the need for temporary intermediate files. Pipes can be used in many instances. For example, the following command would send a long listing of your directory to the printer.

```
$ ls -l | lpr
```

☞ SUMMARY

UNIX is a command-based operating system, and we have used a number of them already in the introduction session. These commands have varied usage and often have a large number of options and need to know more about the general characteristics of commands and the documentation associated with them. In this chapter also introduce all types of commands like as general purpose such as cal, date, who finger etc. Files and directories related commands such as cat, mkdir, rm, cp, etc. Mathematical related commands such as factor, expr, bc, etc. Disk related commands such as df, du, etc. Process and system related commands such as ps, nchup, kill, etc. In this chapter we have learn about the defintion of commands, rules of commands, internal & external commands etc.

☞ EXERCISE

1. Write syntax and uses of the following commands.
   (i) mkdir (ii) passwd (iii) chown
   (iv) umask (v) tee
2. Give the syntax of the following commands and explain the uses also :
   (i) nfs (ii) kill (iii) find
3. Write the syntax and uses of the following commands
   (i) chmod (ii) stty (iii) cat
   (iv) chown (v) In
4. Give the syntax of the following commands and exlain the uses also
   (i) comm (ii) chown (iii) PID
   (iv) diff
5. What is the purpose of permissions in UNIX file system? Explain the different levels of permissions.
6. Write the commands to perform the following functions :
   (i) Display the directory page by page
   (ii) Find out how many users are presently logged.
   (iii) Count the number of words and characters in a file.
   (iv) Delete all files with names starting with 'K'
   (v) List all the processes running on the system.
7. What is the triad of file persmission? Discuss their effect on regular and directory files.
8. Write short not on internal and external commands of UNIX.
9. What are the modes (chmod) of execution in UNIX ? Explain the services of these mode.

❑❑❑

CHAPTER 5

UNIX SHELL PROGRAMMING

5.1 INTRODUCTION

One of the powerful things about UNIX is the extensive use of programmable shells i.e. programs that allow us to run programs and see their output. Shell is the command line interpreter. Users interact with system through shell. The shell interprets the commands that we type, generally acting as the user interface to the UNIX operating system. Many different UNIX shells are available. Shell is command interpreter. It is actually the interface between the user and kernel.

Shell is also called command processor. The shell plays an important role in system that how we interface to the UNIX operating system. The shell provides the interface between us and the UNIX operating system kernel.

The human interface point for UNIX is the *shell,* which is a program or collection of program that provides us with an environment in which to enter commands and parameters to produce a given result.

We have seen that the shell is used to interpret command lines, maintain variables, and execute programs. The shell is also a programming language. We can store a set of shell commands in file. This is known as a **shell script** or **shell programming**. By combining commands and variable assignments with flow control and decision making, we have a powerful programming tool. Using the shell as a programming language, we can automate recurring tasks, write reports, and build and manipulate your own data files. We will discuss shell programming in more detail in next chapter Chapter 6.

A shell is a command line interpreter. It takes commands and executes them. As such, it implements a programming language. The Bourne shell is used to create shell scripts — i.e. programs that are interpreted/executed by the shell.

Even though there are various graphical interfaces available for Linux the shell still is a very neat tool. The shell is not just a collection of commands but a really good programming language.

We can automate a lot of tasks with it, the shell is very good for system administration tasks, we can very quickly try out if your ideas work which makes it very useful for simple prototyping and it is very useful for small utilities that perform some relatively simple tasks where efficiency is less important than ease of configuration, maintenance and portability. The Shell programming is so powerful that several tasks like checking disk space of the system, system shutdown.

5.2 TYPES OF SHELLS

UNIX has provided different shells. These are:

1. The Bourne Shell
2. The C Shell
3. The Korne Shell
4. The Bash (Bourne Shell Again)
5. tcsh
6. zsh
7. EShell The Emacs Shell

1. The Bourne Shell

The Bourne Shell created by the Steve Bourne so it is called Bourne Shell. The Bourne Shell is an important shell and this shell used by many users. The Bourne shell has the following advantages:

- The "standard" AT&T shell — it is found on almost all UNIX systems
- Compact and requires minimal resources
- Executes quickly
- Widely accepted or portable scripts

The file ".profile" located in our home directory is executed each time you login if we are using the Bourne Shell. We can customize our default working environment by placing environmental definitions in this file.

2. The C Shell

C Shell Created by Bill Joy then he is pursuing his graduation at the University of California at Berkeley. This shell is a hit with those who are seriously into UNIX programming. It has two advantages over the Bourne Shell.

(i) **It allows aliasing of commands.** That is we can decide what name we want to call a command.

(ii) **Command line history.** That is C Shell keep track of all commands issued at the command line. Includes the following features C Shell also has many other features :

- Extended Setup Files
  - .login
  - .cshrc (set, history, and alias commands)
  - .logout
- Command Alias
- Command History
- Process (Job) Control
- Script Syntax Similar to C Language

3. The Korn Shell

This shell, written by David Korn, of AT & T Bell Labs, is now the defacto-standard shell for UNIX systems. It's a reasonable platform for both interactive system control as well as shell script programming, providing far superior functionality to the Bourne Shell. It is compatible with and includes most of the features of the Bourne shell. It also includes interactive features analogous to those of the C shell.

4. Bourne Shell-Again Shell-Bash

The FSF folks created an exceedingly souped-up version of the Bourne Shell with automated command completion and plenty of additional functionality. UNIX systems typically choose Bash as the "basic" shell used by default and for managing init.

5. tcsh

Extended version of the C Shell that adds such functionality as command completion, a command line editor, enhanced history manipulation. The "t" is for Tenex; this shell was initially developed on the Tenex operating system.

6. zsh

The zsh is a really powerful shell that provides fairly much more functionality of ksh, csh, and bash thrown together.

One area in which it is dramatically powerful is in its ability to interactively evaluate and modify arguments. It can do command name completion, filename completion, and is extensible such that it can use programs to evaluate possible command completions.

7. EShell – The Emacs Shell

A command shell implemented entirely in Elisp that runs in an Emacs buffer.

Two primary reasons to get excited about this are if:

- You run Emacs everywhere, making EShell more portable than its alternatives;
- You have a great deal of ELisp functions that you would like to be able to use in scripts.

Summary of shell facilities

	Bourne	C	TC	Korn	BASH
command history	No	Yes	Yes	Yes	Yes
command alias	No	Yes	Yes	Yes	Yes
shell scripts	Yes	Yes	Yes	Yes	Yes
filename completion	No	Yes*	Yes	Yes*	Yes
command line editing	No	No	Yes	Yes*	Yes
job control	No	Yes	Yes	Yes	Yes

* not the default setting for this shell.

5.2.1. Features of UNIX Shell

It doesn't matter which of the standard shells you choose, because they all have the same purpose: to provide a user interface to UNIX. To provide this interface, all the shells offer the same basic characteristics:

1. Shell meta-characters (wild cards)

2. Maintenance of variables
3. File handling: input/output redirection and pipes
4. Access to and handling of program commands
5. Special Command Line Characters
6. Environment control
7. Reserved words
8. Shell programming
9. Command-line interpretation

5.3 SHELL META-CHARACTER (WILDCARDS)

All shell versions have meta-characters, which allow the user to specify filenames. The shell gives us a way to reduce by contraction filenames through the use of special patterns called wildcards. We can use wildcards to specify a whole set of files at once, or to search for a file when we know only some or its name. The following are wild cards:

Wild Card	Description
*	Matches any portion
?	Matches any single character
[]	Matches a range or list of characters
;	seperates multiple commands on the same line
&	runs the preceding command in the background
/	nullifies the special meaning of the next metacharacter

Wildcard characters provide convenient shorthand for specifying multiple file or directory names with a single name. Two of the most useful wildcard characters are the * and the?. The * matches any arbitrary character string (including the null character), and the ? matches any single character.

5.3.1 The * Wildcard

The * wildcard means 'any characters, including no characters'. For example, suppose we have created the following files in our current working directory:

```
$ls
myfile myfile2 myfile3 test yourfile
```

To list only the file names beginning with 'myfile', we could type

```
$ls myfile*
myfile myfile2 myfile3
```

Even though test and yourfile exist, ls displayed only the file names that start with 'myfile'. If we wanted to list the file names containing 'file', we could type:

```
$ls *file*
myfile myfile2 myfile3 yourfile
```

5.3.2 The ? Wildcard

The ? wildcard means 'any single characters'. Although you probably will not use this wildcard as much as the *, it is still useful. For example, if we want to list only the files that start with 'myfile' and end with a single additional character, we could type:

```
$ls myfile?
myfile2 myfile3
```

The ? wildcard character matches exactly one character. Thus, 'myfile' did not show up in this listing because it did not have another character at the end. Wildcards can be used with any Unix command which takes files or directories as parameters.

Using the * wildcard characters with mv, cp, and rm

Wildcard characters are often useful when you want to move or copy multiple files from one directory to another. For example, suppose you have two directories immediately below your current directory, named 'new' and 'old', and these directories contain the following files:

```
$ls new
myfile myfile2
$ls old
myfile3 myfile4
```

To move all the files from the directory 'new' into the directory 'old', we would type:

```
$ mv new/* old
$ls new ( no files in new)
$ls old ( all files are now in old)
myfile myfile2 myfile3 myfile4
```

We can do a similar operation with the cp command. For example, to copy all the files from old into new, we could type:

```
$cp old/* new
```

Similarly, we can use wildcard characters with the rm command. For example, to remove all of the files in the directory new, type:

```
$ rm new/*
$ls new ( all the files are now gone )
```

Warning: Be careful when using wildcard that you do not accidentally remove or move files that you need.

5.3.3 The Bracket ([ ]) Wildcard Character

The third type of wildcard is actually a pair of square bracket characters. This wildcard is an expansion matching any single character or range of characters separated with a dash ('-') listed within the brackets. For example

```
$ls *[2-4]
myfile2 myfile3 myfile4
or if you wish to be more specific:
$ls myfile[2-4]
```

```
myfile2 myfile3 myfile4
```

Notice the - character specifies a range. If you only wanted to list files that included the numbers 2 and 4 then you could type:

```
$ls myfile[24]
myfile2 myfile4
```

Wildcards can be useful when processing a number of specific files. The following are some examples:

```
$ls t*
```

This lists all files starting with t.

```
$ls test?5.dat
```

This lists all files starting with test, any single character and ends with 5.dat.

```
$ls [a-c]*
```

This lists all files starting with a through c.

```
$ls [e,m,t]*
```

This lists all files starting with e, m, or t.

5.3.4 Quoting Metacharacters

The following characters come under this:

" " , \ , ` ` , ' ',

5.4 SHELL VARIABLES

In programming, a variable is a name associated with a value. It offers a symbolic way to represent and manipulate data. Shell variables come in two varieties: those created and maintained by the UNIX system itself, and those created by user.

Shell variables are really just the general case of environment variables. If we are a programmer, remember that a UNIX shell really runs an interpreted programming language. Shell variables belong to the shell; we can set them, print them, and work with them much as we can in a C program or other program. If we are not a programmer, just remember that shell variables are pigeonholes that store information for you or your shell to use.

The rule for building shell variables in UNIX are as follow:

(a) 1-20 characters (alphabetic, numeric, or an underscore).

(b) A variable name is any combination of digits, alphabets and an underscore.

(c) No commas or blanks space are allowed within a variable name.

(d) The first character of a variable name must be an alphabet or an underscore.

(e) Variables names should be of any reasonable length.

(f) Variable names are case-sensitive i.e., Name, name, Name and NAME are all different variables.

By convention, shell variables have lowercase names. The shell variables are of two types:

1. User- define variable.
2. System shell variable.

1. User-Define Variable

To create the shell variable by user, simply type in the form which is given below

```
Name=value
```

A shell variable is created as name in the left hand and having the right hand value.

For example,

```
fruit=mango
```

Create a shell variable having the name fruit and its value is string mango. Shell variable by default string variable. We can convert string value into digit string by using expr command.

For example

```
Length=125
```

Here the value of length is a three character string consisting of the digit characters "1", "2" , and "5". The value of weight is not the number 125.

Note: No space are used when we assign a value to a variable.

For example

```
fruit = mango
```

This value will not to be accepted.

To find the value of a shell variable, we can use the echo command:

```
$echo fruit
fruit
$
```

If we use a variable name preceded by a $ as an argument to echo

```
$echo $fruit
mango
$
```

Here fruit is the variable's name and $fruit is the variable value. We can use both in the same echo statement:

```
$echo The value of fruit is $fruit
```

The value of fruit is mango.

2. System Shell Variable

The shell maintain its own set of shell variables are called system shell variables. But to see what your system is using, just type set. Here is a typical response:

```
$set
EXINIT='set ai nu'
HOME=/usr/fafner
IFS=
MAIL=/usr/mail/fafner
PATH=:/bin:/usr/bin
PS1=$
PS2=>
```

```
TERM=adm5
$
```

5.4.1 Some Standard Shell Variables (Environment Variables)

When a process begins execution, one of the exec family of functions makes available an array of strings called the environment. Here are some standard shell variables found on many systems. Which are given below:

EXINIT: Initialization instruction for vi and ed editors.

HOME : This variable is interpreted as the pathname of the user's home directory.

IFS (Internal Field Separators): A string treated as a list of characters that is used for field splitting and to split lines into fields with the read command. This is set to a list of the characters that are used to separate words in a command line.

LOGNAME or USER : LOGNAME or USER the userid of the person who is currently logged in.

LANG : Provide a default value for the internationalization variables that are unset or null.

PATH : PATH the list of directories to be searched by the shell to find commands that use by us. This variable represents a string formatted, used to effect command interpretation.

PPID : This variable is set by the shell to the decimal process ID of the process that invoked this shell. In a subshell, PPID will be set to the same value as that of the parent of the current shell.

PS1 (Prompt String 1) : Each time an interactive shell is ready to read a command, the value of this variable is subjected to parameter expansion and written to standard error. The default value is "$". For users who have specific additional implementation-dependent privileges, the default may be another, implementation-dependent, value.

PS2 (Prompt String 2) : Each time the user enters a newline character prior to completing a command line in an interactive shell, the value of this variable is subjected to parameter expansion and written to standard error. The default value is ">".

PS4 (Prompt String 4) : When an execution trace (set **-x**) is being performed in an interactive shell, before each line in the execution trace, the value of this variable is subjected to parameter expansion and written to standard error. The default value is "+".

PWD : OWD the absolute pathname of current directory.

TERM : TERM the type of terminal being used.This identifies the kind of terminal we habitually use.

EDITOR : EDITOR the pathname of the default text editor.

5.4.2 Special Shell Variables

There are some variables which are set internally by the shell and which are available to the user:

Name	Description
$1 - $9	these variables are the positional parameters.
$0	the name of the command currently being executed.
$#	the number of positional arguments given to this invocation of the shell.
$?	the exit status of the last command executed is given as a decimal string. When a command completes successfully, it returns the exit status of 0 (zero), otherwise it returns a non-zero exit status.
$$	the process number of this shell - useful for including in filenames, to make them unique.
$!	the process id of the last command run in the background.
$-	the current options supplied to this invocation of the shell.
$*	a string containing all the arguments to the shell, starting at $1.
$@	same as above, except when quoted.

5.5 SHELL SCRIPT

There are a lot of different shells available for UNIX/Linux but usually the sh (Bourne shell) is used for shell programming as it is available for free and is easy to use. So all the scripts we will write in this shell i.e. sh. For writing our shell programs we can use any kind of text editor, e.g. emacs, vi,ed etc. as with other programming languages.

As we know that a user can not interact with the kernel directly. The shell programming skills are a must to be able to exploit the power of UNIX to the complete extent.

The shell programs can be used in Shell programming for a variety of tasks. This is given below:

(a) **Customizing our work environment:** For example, if we want to display the current date and time then we can write the shell script for this purpose.

(b) **Automating our daily tasks:** For example if we want to back up all our programs at the end of every day. This is possible by using shell script.

(c) **Automating respective tasks:** For example if we want the sales report of the business every month. This can be done using shell script.

(d) Executing important system procedures like formatting the disk, creating file system, shutting down the system.

When you have written your script and saved it you have to make it executable to be able to use it.

To make a script executable type

```
$chmod +x filename
```

5.5.1 Read Statement

Use to get input from keyboard and store to variable :

Syntax of read statement is :

```
read variable1, variable2, .... variable n
```

5.5.2 Simple Shell Script

```
$vi anu
>echo enter the name
>read name
>echo "U enter the name $name"
```

Output

```
Enter the name
Anita
U enter the name Anita
```

5.6 SHELL COMMANDS

Most UNIX commands are separate commands run by the shell, but commands are part of the shell (sh) program itself. Since no new process need be started upto run these built-in commands because they use less time than the regular commands. These commands are called shell commands, such as set, shift, etc.

5.6.1 Positional Parameters: Set

Some UNIX commands like as data produce single-line-output. We also pass command output through filter like tail & head to produce a single line. This is overkill, the shell has an internal command to do this job. The set, set statement assigns its arguments to the positional parameter $1, $2, $3 and so on. Let us set to convert its arguments to positional parameters :

```
$set Hi Any How are you
$
```

This assigns the value Hi to the positional parameter $, Anu to $2, $3 to How, $4 to are, and $5 to you. It is also sets the other parametere $# and $*.

```
$echo "the $# arguments are $*".
```

The 5 arguments are Hi Anu How are you.

We can display the particular positional parameter like this

```
$echo $1
Hi
$ech $1 $2
Hi Anu
```

5.6.2 Shifting Arguments: Shift

Shift transfers the contents of a positional parameters to its immediate lower numbered one. This is done as many times as the statement is called. When called once, $2 becomes $1, $3 become $2 and so on.

```
$echo $1 $2 $3 $4 $5
      Hi Anu How are you
```

```
$shift 2
$echo $1 $2 $3 $4 $5
    How are you
```

5.6.3 Changing Your Login Shell

In some cases, you may want to change your login shell. You may need a specific feature that your default shell locks.

For example, sh and csh lock some very important features of the newer shells, and tcsh is often criticized for not being compatible witgh Bourne shell scripts.

The command to change your login shell is chsh. In this example, the user tanmay is changing his shell to bash :

```
$ chsh
  changing shell for tanmay
  Password :
  New shell [bin/csh] : /bin/bash
  Shell change
$
```

As you can see, tanmay will need to enter his password before he can change his shell. His current shell is/bin/csh. To change it. he needs to know that full path name for the new shell.

To find the pathname for a command or executable, type which followed by the name of the command.

For example,

```
$ which bash
/ bin / bash
```

You can also find the full pathnames for the shells installed on your system in the / etc/ shells file.

5.7 THE ENVIRONMENT

When a shell variable is exported, it is added to the environment. The environment is the set of all exported variables. The importance of the environment is that a process can access and use any of the variables in its environment.

Here are some ways of the environment :

- Use the export command to add a new variable to the list.

  For example:

  $name = kumar

  $export kumar

- Modify the value of an export variable before running a program by typing in the new definition :

  $PATH =/usr/abc/anu

  $MCA

Since the value of the PATH is exported, the MCA program will search only the /usr/abc/anu directory for command.

- Modify an exported variable within a shell script.

  For example, we may wish to have a line like

  $PATH = /bin:/bin/user

- Make a temporary addition to or modification of the environment.

  For example,

  $name = Value command

  This causes the variable definition to hold only for the duration of the command.

  For example,

  $PATH = /user/abc/anu MCA

  PATH to be defined as shown only while the MCA command is run. The PATH assume its original value.

  Let the command is a shell script:

  $echo $xyz

  Plcae it in the executable file xyz1, then we can do the following:

  $xyz = MCA;export xyz

  $xyz1

  MCA

  $echo $xyz

  MCA

5.8 INTEGER ARITHMETIC

Arithmetic expressions are evaluated inside double quotes, like tildes, variables, and command substitutions. We are finally in a position to state the definitive rule about quoting strings: When in doubt, enclose a string in single quotes, unless it contains tildes or any expression involving a dollar sign, in which case you should use double quotes.

For example, the date command on System V-derived versions of UNIX accepts arguments that tell it how to format its output. The argument +%j tells it to print the day of the year, i.e., the number of days since December 31st of the previous year.

We can use +%j to print a little holiday anticipation message:

```
print "Only $(( (365-$(date +%j)) / 7 )) weeks until the New Year!"
```

The arithmetic expression feature is built in to the Korn shell's syntax, and was available in the Bourne shell only through the external command expr. Thus it is yet another example of a desirable feature provided by an external command being better integrated into the shell.

Korn shell arithmetic expressions are equivalent to their counterparts in the C language. Precedence and associativity are the same as in C. Table 5.1 shows the arithmetic operators that are supported. Although some of these are special characters, there is no need to backslash-escape them, because they are within the $((...)) syntax.

For example, $((x += 2)) adds 2 to x and stores the result back in x.

Table 5.1 : Arithmetic Operators

<table>
<tr><th>Operator</th><th>Meaning</th></tr>
<tr><td>+</td><td>Plus</td></tr>
<tr><td>-</td><td>Minus</td></tr>
<tr><td>*</td><td>Times</td></tr>
<tr><td>/</td><td>Division (with truncation)</td></tr>
<tr><td>%</td><td>Remainder</td></tr>
<tr><td><<</td><td>Bit-shift left</td></tr>
<tr><td>>></td><td>Bit-shift right</td></tr>
<tr><td>&</td><td>Bitwise and</td></tr>
<tr><td>|</td><td>Bitwise or</td></tr>
<tr><td>~</td><td>Bitwise not</td></tr>
<tr><td>^</td><td>Bitwise exclusive or</td></tr>
</table>

Parentheses can be used to group sub expressions. The arithmetic expression syntax also supports relational operators as truth values of 1 for true and 0 for false. Table 5.2 shows the relational operators and the logical operators that can be used to combine relational expressions.

Table 5.2 : Relational Operators

<table>
<tr><th>Operator</th><th>Meaning</th></tr>
<tr><td><</td><td>Less than</td></tr>
<tr><td>></td><td>Greater than</td></tr>
<tr><td><=</td><td>Less than or equal</td></tr>
<tr><td>>=</td><td>Greater than or equal</td></tr>
<tr><td>==</td><td>Equal</td></tr>
<tr><td>!=</td><td>Not equal</td></tr>
<tr><td>&&</td><td>Logical and</td></tr>
<tr><td>||</td><td>Logical or</td></tr>
</table>

For example,

$((3 > 2)) has the value 1; $(((3 > 2) || (4 <= 1))) also has the value 1, since at least one of the two sub expressions is true.

5.8.1 Arithmetic Expressions

The Korn shell's let command performs integer arithmetic. The Korn shell provides a way to substitute integer values; base conversion is also possible :

$((expr)) : Use the value of the enclosed arithmetic expression.

B # n : Interpret integer n in numeric base B.

For example, 8#100 specifies the octal equivalent of decimal 64.

Examples

See the let command for more information and examples.

```
let "count=0" "i = i + 1"
Assign
   i
   and
   count
   . let "num % 2"
Test for an even number.
      ((percent >= 0 && percent <= 100))
Test the range of a value.
```

5.9 STRINGS MANIPULATING

Bash supports a surprising number of string manipulation operations. Unfortunately, these tools lack a unified focus. Some are a subset of parameter substitution, and others fall under the functionality of the UNIX expr command. This results in inconsistent command syntax and overlap of functionality, not to mention confusion.

String Length

```
${#string}
expr length $string
```

These are the equivalent of strlen() in C.

```
expr    "$string" : '.*'
stringZ=abcABC123ABCabc
echo    ${#stringZ}                    # 15
echo    `expr length $stringZ`         # 15
echo    `expr "$stringZ" : '.*'`       # 15
```

Length of Matching Substring at Beginning of String

```
expr match "$string" '$substring'
      $substring is a regular expression.
expr "$string" : '$substring'
      $substring is a regular expression.
stringZ=abcABC123ABCabc
#     |------|
#     12345678
echo   `expr match "$stringZ" 'abc[A-Z]*.2'`      # 8
echo   `expr "$stringZ" : 'abc[A-Z]*.2'`          # 8
```

Index

```
expr index $string $substring
```

Numerical position in $string of first character in $substring that matches.

```
stringZ=abcABC123ABCabc
#  123456 ...
echo `expr index "$stringZ" C12`    # 6
                                    # C position.
echo `expr index "$stringZ" 1c`     # 3
#  'c' (in #3 position) matches before '1'.
```

This is the near equivalent of strchr() in C.

Substring Extraction

```
${string:position}
```

Extracts substring from $string at $position.

If the $string parameter is "*" or "@", then this extracts the positional parameters, starting at $position.

```
${string:position:length}
```

Extracts $length characters of substring from $string at $position.

5.10 SPECIAL COMMAND LINE CHARACTERS

Process Execution Metcharacters or Sequential Command

When we want to run more then one command at the same time and same command prompt($) in one line, then we separate them with semicolon.

For example,

```
$ls;pwd;who
```

This would result in the execution of ls first, then pwd, and lastly the who command display all login users.

Another example:

```
$ls -l anu; mv anu arya; ls -l arya
```

5.10.1 Conditional Execution Using && and ||

These special command line character can be used to optionally execute a command depending on the failure or success of the previous command

If we want the second command to be execute only if the first command succeeds:

```
$command1 && command2
```

If we want the second command to be carries out only if the first fails, shell provides the special command line character ||

```
$command1 || command2
```

For example:

```
$grep MCA anu && echo save anu
```

Only if the string were found in the file anu.

```
$grep BCA anu || echo no BCA
```

The special command line character || performed when the initial condition is failed.

5.10.2 Command Grouping: (Parentheses)

We can use parentheses to group commands.

For example: We wish to send the current date and time along with the content of a file called anu to file arya

```
$(date;cat anu) >arya
```

5.10.3 Inserting Comments

Another useful special command line character or metacharater in shell script # (hesh symbol). It tells the shell to ignore what follows.

For example:

```
$# this is UNIX Operating system.
```

5.10.4 Double Quotes

Anything enclosed in double quotes is passed on exactly as presented with the exception that the values of shell variables are substituted.

For example

```
v1="A B C"
v2="$v1 D"
echo $v1 $v2
```

This will produce the output

```
A B C A B C D
```

Command substitution is also performed for double quote enclosed strings.

5.10.5 Single Quotes

Any digit or alphabet enclosed in single quotes is passed on exactly as presented. The values of shell variables are not substituted.

For example

```
v1="a b c"
v2='$v1 d'
echo $v1 $v2
```

This will produce the output

```
a b c $v1 d
```

5.10.6 Back Quotes

Back quotes are used to enclose commands. An item enclosed in back quotes is replaced by the standard output of the command. Shell variable values are substituted within back quotes.

For example

```
$date='date'
```

```
$echo the date is $date
```

Producing the output

```
the date is Sat 30 Dec 10.27 IST 2007
```

A further example is shown below

```
FILENAME=$HOME/.profile
LINES='wc -l < $FILENAME'
echo There are $LINES lines in the file $FILENAME
```

This produced the output

```
There are 8 lines in the file /home/staff/acad/jphb/.profile
```

5.10.7 Back Slash

The back slash symbol may be used to turn-off the special effects of any single character after the back slash.

For example

```
v3=a\ b
echo $v3
```

5.10.8 Producing the Output

a b

It is important to note that strings may be presented to the shell more than once, in which case it may be necessary to use a \ to protect (or **escape**) the quotes that are going to be significant second time.

For example.

```
msg='echo Your Current Directory is \'pwd\"
```

To evaluate the string the *echo* command is issued with the argument list

```
Your Current Directory is 'pwd'
```

i.e., the escaped backquotes around *pwd* are presented literally as backquotes to the *echo* command. The effect of all this is to set the shell variable *msg* to a suitable string of characters,

For example

```
Your Current Directory is /home/staff/acad/jphb/modules/ospp
```

5.11 ADVANTAGE AND DISADVANTAGE OF SHELL SCRIPT

5.11.1 Advantages

1. Writing a shell script is much quicker than writing the equivalent code in other programming or scripting languages. As with other interpreted languages, shell scripts have no compilation step, so the script can be executed quickly while debugging.
2. It has local and global variables
3. Control instructions, functions etc.
4. Ix we are to execute a shell program we don't need a separate compiler .

5.11.2 Disadvantages

Using shell scripts can also have a number of significant disadvantages.

1. One significant disadvantage of using shell scripts is that they can run slowly due to the need to create potentially many new sub-processes for each of the many commands executed. When a script's job can be accomplished by setting up a pipeline in which efficient filter commands perform most of the work, the slowdown is minimal. But if a shell script has to perform multiple individual actions on many individual data items, resulting in multiple forks for each command invoked to perform each action, the script might be orders of magnitude slower than a conventional compiled program (in which those same actions might require but single processor instructions).
2. Simple sh scripts can be quite compatible with the extremely diverse range of Unix, Linux, and BSD operating systems, versions thereof, and system utilities—but more complex shell scripts can be fail because of the many differences between shells, utilities, and other crucial elements. Larry Wall of Perl famously wrote that "It is easier to port a shell than a shell script."
3. Similarly, more complex scripts can be run into the limitations of the shell scripting language itself; the limits make it difficult to write quality code and extensions by various shells to ameliorate problems with the original shell language can make problems worse.

5.12 DECISION MAKING

Decision-making is also called control structure. Control structure in which the various steps are executed sequentially. Most of the time we reed the program. To do one thing under same circumstances and a different thing in another situation.

The control structures are as follow :

1. The if-then-fi statement
2. The if-then-else-fi statement
3. The if-then-elif-else-fi statement
4. The case-esac statement.

5.12.1 The if-then- fi statement

The general form of if statement is

```
if control command
command 1
fi
```

The if-then-fi statement is a powerful decision making statement and is used to control the flow of execution of commands.

The command may be a single command or a group of commands. If the control command is true, the command / will be execute; otherwise there will be no operation, i.e. exit the statement.

For example

```
    $ vi copy file
    echo enter source file and target file names
    read source target
    if cp $ source $ target
    then
    echo file copied successfully
fi
```

In this script first we read two files, source file and target file and then command cp proceeds to copy the contents of source file into target file. If it is able to do so successfully it would return file copies successfully otherwise exit.

Flow chart of if-then fi statement

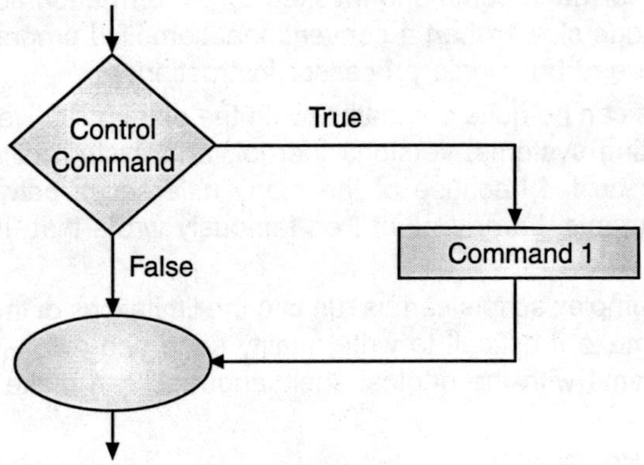

FIGURE 5.1.

5.12.2 The if-then-else-fi-statement

The if-then-else-fi statement is an extension of if-then-fi statement.

The general form of If-then-else-fi statement is

```
if control command
then
   command 1
else
   command 2
fi
```

if the control command is true, then the command 1 will be executed, otherwise, the command 2 will be executed.

For example

```
$ vi cp file
   echo enter source file and target file
read source target
   if Cp $ source $ target
then
   echo file copied successfully
else
   echo copy failed
fi
```

If control command executes successfully, then the echo between then and else gets executed, otherwise the echo between else and fi gets executed.

Flow chart of if-then-else-fi statements

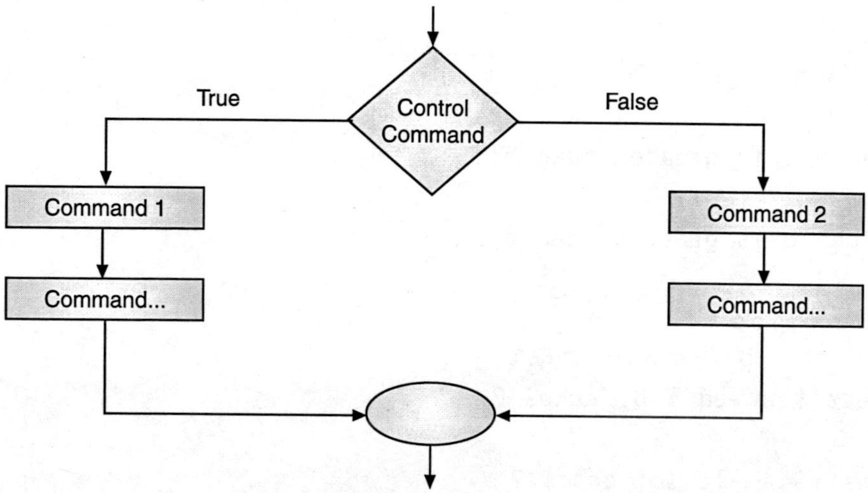

FIGURE 5.2.

The if statement can be represent by a block diagram as given above.

The group of commands between the then and else is called an 'if block.' Similarly, the commands between the else and fi form the 'else block'

The Test command

We may decide to execute a set of commands if the first argument supplied to the script is a file. Or in other case, we may execute a set of commands if the number of arguments is less than four. To take care of such situation unix provides a command called test.

The test command can be used to evaluate logical expressions in our if statements. When test evaluates a true expression, it returns O. it the expression is false, the test returns a non-zero value. The test command can carry out several types of test. These are given below:

1. Numerical tests.
2. String tests.
3. File tests.

1. Numerical Tests

Numerical comparison operators used by test

Operator	Meaning
-eq	equal to
-ne	not equal to
-gt	Greater than
-ge	Greater than or equal to
-lt	Less than
-le	Less than or equal to

For example greatest number out of two numbers

```
$ vi greatest
echo enter two number
read a b
if [ $ a - gt $ b]
then
   echo a is greater than b
elso
   echo b is greater than a
fi
```

Another example

```
$ a = 5;  b = 8;  c = 8.1
$ test $ a -eq $ b; echo$ ?
1
$ test $ a -lt $b; echo$ ?
0
$ test $c -gt $b; echo $ ?
1
$ test $c -eq $b; echo$?
0
```

2. String Tests

We can also use test command to examine strings. The test command can handle the string tests.

String comparison operators used by test are :

Operator / condition	Meaning
-z string	Length of string is zero
- r string	Length of string is non zero
string	string is not the null (same-n)
string 1 = string 2	string 1 is equal to string 2
String 1 ! = String 2	string is not equal to string or strings are not equal

For example:

```
# String test
string 1 = "Morning"
string 2 = "Even"
string 3 =
[ $ string 1 = $ string 2]
echo$?
```

```
[ $ string 1! = string 2]
echo $?
[ - r$ string1]
echo $ ?
[-z'' $ string 3'']
echo$?
[ -z $ string 3]
echo $?
[ '' $ string 3'']
echo $?
```

Here is the out put of the program

```
1
0
0
0
0
```

Expected output of test command for string comparison.

3. File Tests

The test command has several options for checking the status of a file. The options are given below:

Options	Meaning
-a file	file exists
-r file	file exists and is readable
-w file	file exists and is writable
-x file	file exists and is a append file
-f file	file exists and is a regular file
-d file	file exists and is a directory
-h file	file exists and is a symbolic link
-c file	file exists and is a file.
-s file	file exists and has a size greater than zero
-k file	file exists and its sticky bit is set

Some example or script for file testing

```
$ vi file test
   echo Enter file name
read frame
if  [- f $ frame]
then
```

```
    echo The given file is a regural file
else
    echo this is not a file
fi
```

Another example for checking whether the user has a write permission to a file. If yes, it prompts the user to type a message which then gets appended to mentioned file.

```
$ vi file testing
    echo enter file name
    read w file
if [- w $ wfile]
then
    echo we can append for save and stop type ctrl d
    cat >> $ w file
else
    echo no write permission
fi
```

By using redirection operator (>>) gets appended to the desired file. When we hits ctrl+D the cat command comes to an end.

5.12.3 The if-then-elif-else-fi statement

There is another way of putting its together when multipath decisions are involved. A multipath decision is a chain of its in which the command associated with each else has an if. If we write an entire if-else-fi construct within either the body of the if statement or the body of an else statement. This is called nested if-else.

The general form of nested if-else given below:

```
if control command 1
then
    command 1
else if control command 2
then
    command 2
else if control command 3
    then
    command 3
        :
        .
    else if control command n
    then
    command n
```

```
else
    default
fi
```

Let us consider an example of grading the students of an institute. The grading of student will be done according to the following rules:

Avg marks	Grade
80 to 100	Honours
60 to 79	First division
45 to 59	Second division
30 to 44	Third division
0 to 29	Fail

```
$ vi grade
echo enter the average marks
read marks
if [ $ marks - gt 79]
then
    echo Grade = Honours
else if [ $ marks -gt 59]
    then
    echo Grade = first division
else if [ $ marks - gt 44]
  then
    echo Grade = "Second Division
else if [ $ marks - gt 29]
    then
    echo Grade = Third Division
else
  echo Grade = fail
fi
fi
fi
fi
```

Flow chart of nested if-else statement

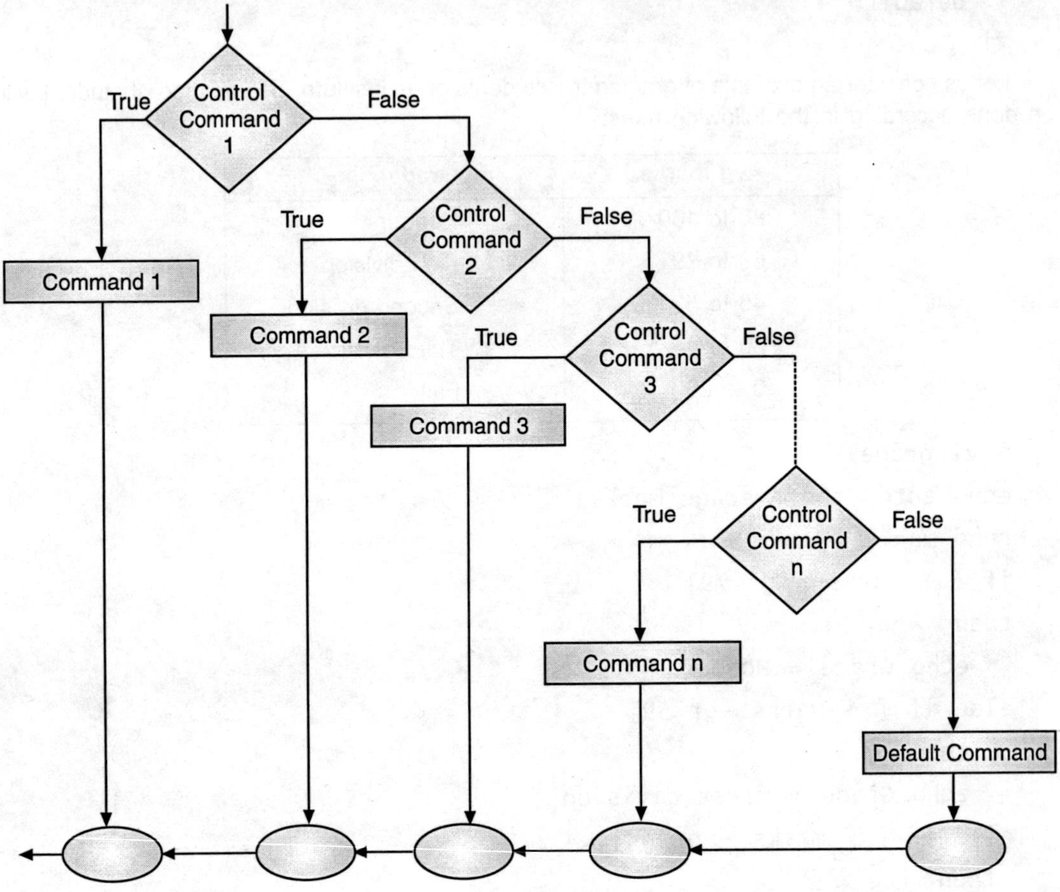

FIGURE 5.3.

The second if-else construct is nested in the first else statement. If the test in the first if statement fails, than the test in the second if statement is performed similarly third and so up to n if statement is false then the final else statement is executed .

Logical operators

In unix shell allows usage of three logical operators. These are:

1. -a (AND)
2. -O (OR)
3. ! (NOT)

The first two operators -o and -a may be use as a combination in a test.

For example, to generate a marksheet of a student with five subjects. The student gets a division as for the following rules:

Percentage	Division
60 to 100	First
45 to 50	second
30 to 44	Third
0 to 29	Fail

```
$ vi marksheet
echo enter the marks of five subject
read m1 m2 m3 m4 m5
$ = ‘expr $m1+ m2 + m3 + $m4 + $m5’
   per = ‘expr $ s/5’
if [ per -ge 60]
then
   echo First Devision
fi
if [ per - ge 45 - a per - lt 60]
   then
   echo Second Division
fi
if [ per-ge 30 - a per lt 45 ]
   then
 echo Third Division
fi
if [ per lt 29]
   then
   echo Fail
fi
```

In the given script the—a operator is used to combine two conditions. If both the conditions is true, we get result printed as second division. It one the conditions evaluate to false then the whole thing is treated as false.

The third logical operator is the NOT operator, written as!. This Operator reverses the values of the expression on which it operates, it makes a true expression false and a false expression true.

For example, test a file that it is directory or not.

```
$ vi d file
echo enter file name
reat d name
if [ ! - d name]
   then
   echo this is a directory
else
   echo this is not a directory
fi
```

5.12.4 The case- esac Statement

The case-esac statement is the second conditional offered by the shell. The statement maches an expression for more than one alternative, and uses a compact construct to permit multi-way branching.

The general form of the case-esac statement given below:

```
Case value in
Choice1)
   expression 1
   ;;
Choice2)
   expression 2
   ;;
Choice3)
   expression 3
   ;;
*) default
   ;;
   esac
```

The expression in the above represents any valid unix command. Also, choice1, choice2, and choice3, are labels which identify the potential choices of actions.

Flow chart of case-esac statement is given below:

case $ choice in

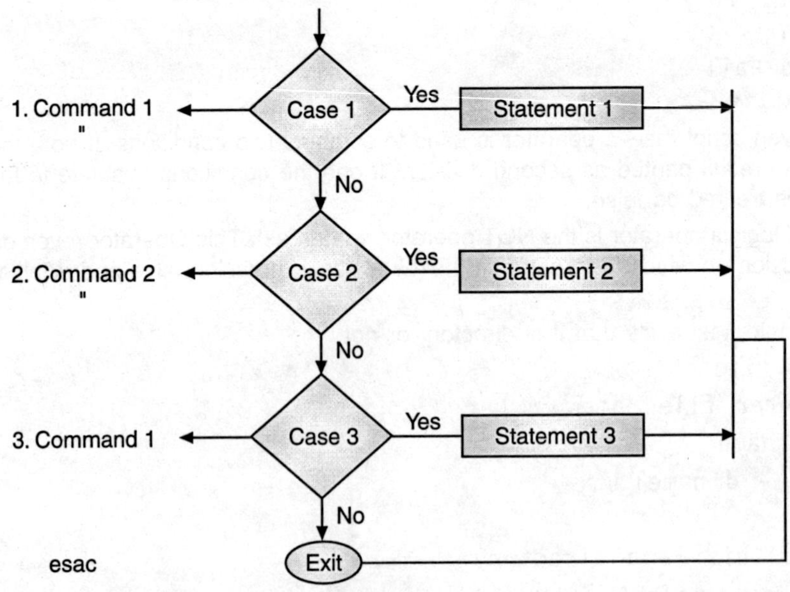

FIGURE 5.4.

For example, consider a simple shell script menu, that uses case. The script accepts vaules from 1 to 5 and performing some action depending on the enter keys.

We can see the list of files, current date, user of the system etc.

```
$ vi menu
echo main menu
echo 1 list of file
echo 2 current date
echo 3 user of the system
echo 4 calender
echo 5 exit
echo enter the choice
read ch
case "$ ch" in
1. ls"
2. date"
3. who"
4. cal"
5. exit"
6. echo wrong choice"
esac
when we execute this script enter the choice
3
Tue   Jan   1  10: 30 : 06  IST 2008
```

Another example of case statement is given below:

```
$ vi number file
echo '' enter the number 1 to 5
read numb
case $ numb in
1. echo you entered no.1
   "
2. echo you entered no.2
   "
3. echo you entered no.3
   "
4. echo you entered no.4
   "
5. echo you entered no.5
   "
```

```
6. echo wrong entery"
esac
```

5.13 LOOPING CONTROL

A sequence of statements are executed until some conditions for termination of the loop are satisfied, A program loop consists of two segments, are as the body of the loop and the other as the control statement. The control statement tests certain conditions and then directs the repeated of the statement contained in the body of the loop.

A looping process includes the following four steps:

1. Initialization and setting of a counter.
2. Exsection of the statements in the loop.
3. Test for a specified condition for execution of the loop.
4. Incrementing the counter.

There are three loops by way of which we can repeat a part of a program. These are given below:

1. The for loop
2. The while loop
3. The until loop

5.13.1 The for loop

The for loop executes a block of commands once for each member of a list.

The general form of for loop is given below:

```
for variable in value1 value2 value3 ...........
do
    commands
done
```

The variable in the general form can have any name that we choose. We can use for loops to repeat a command a fixes number of times.

For example, if we enter the following on the command line,

```
$ for a in 1 2 3 4 5 6 7 8 9
do
    echo $a
    done
output
1
2
3
4
```

```
5
6
7
8
9
```

Another example of for loop

```
$ for ch in unix is an operating system
do
   echo $ ch
done
out put
unix
is
an
operating
system
```

We can use wild characters in the for loop in case of more than one similar character or words.

5.13.2 The While Loop

The while loop respects a block of commands based on the result of a logical test. The general form of while loop is given below:

```
while test command
do
   commands
done
```

Flow chart of while loop is as follow:

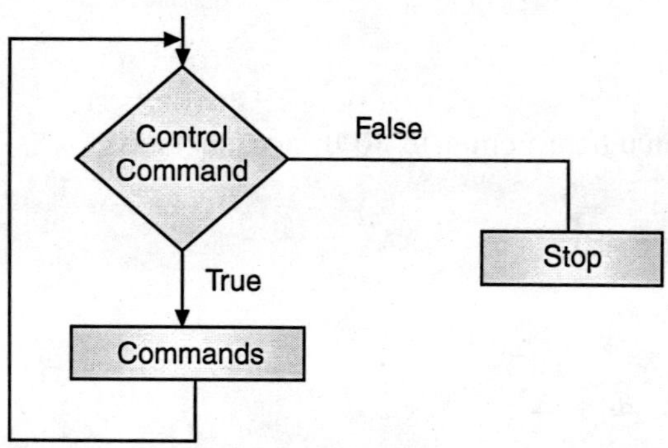

FIGURE 5.5.

When while loop is executed, it runs commands, if the return value of the test command is true, commands are executed, and the program returns to the while test. The loop continues until the value of test command is false, at which point while terminates.

For example, printers the value of integer 1 to 10

```
a = 1
while [ $ a - le 10]
do
   echo $a
   a = ‘ expr $ a + 1
   done
Genorate a table of a given number.
$ vi table
echo enter the no. for table
read r
i = 1
while [$ i -le 10]
   do
   a = ‘expr i \ n’
   i = ‘ expr $i + i’
done
```

5.13.3 The Until Loop

The until loop is the complement of the while loop, and its formand uses are similar. The difference between them is that while loops repeat until the test is false, and until loops repeat until the test is true.

The general form of the until loop is given below :

```
until control command
do
   commands
done
```

For example, prints numbers from 1 to 10 using until loop

```
 a = 1
until [ $ a - gt 10]
   do
   echo $ a
   a = ‘ expr $ a + 1’
done
```

The statements within the until loop may be a single or a group of commands. The until loop executes at least one time, if test condition is true or false in both condition.

5.13.4 The Break Statement

Where we want to jump out of a loop immediately, without waiting to get back to the control-command. The keyword break use us to do this.

We often come across situations where we want to jump out of a loop instantly, without waiting to get back to the control command. The keyword break allows us to do this. When the keyboard break is encountered inside any loop, control automatically passes to the first statement after the loop. A break is usually associated with an if.

For example

```
i = 1 j = 1
while [$1 -le 1]
do
   while [$j -lt 150]
   do
      if [$j -eq 50]
      then
         break
      else
         echo $i $j
      fi
   j = 1 expr $j + 1'
   done
      i = 'expr $i + 1'
   done
```

5.13.5 The Continue Statement

We want to take the control to the beginning of the loop, by passing the statements inside the loop which have not yet been executed. The keyword continue allows us to do this. When the keyword continue is encountered inside any loop, control automatically passes to the beginning of the loop. A continue is usually associated with an if.

For example

```
i = 1
while [$i -le 2]
do
j = 1
while [$j -le 2]
do
   if [$i -eq $j]
   then
      j = 'expr $j + 1'
      continue
fi
```

```
        j = 'expr #j + 1'
        echo $i $j
    done
        i = 'expr $i + 1'
    done
```

5.14 SHELL OPERATORS

There are various operators supported by each shell. But we are discussing only default shell (Bourne) so we are going to cover all the important Bourne Shell operators.

There are following operators :

1. Arithmetic Operators
2. Relational Operators
3. Boolean Operators
4. String Operators
5. File Test Operators

The Bourne shell didn't originally have any mechanism to perform simple arithmetic but it uses external programs, either awk or the simpler program expr.

For example to add two numbers :

```
$/bin/sh
val=`expr 2 + 2`
echo "Total value : $val"
Output:
Total value : 4
```

There are following points to note down :

- There must be spaces between operators and expressions for example 2 + 2 is not correct, where as it should be written as 2 + 2.
- Complete expression should be enclosed between ``, called inverted commas.

5.14.1 Arithmetic Operators

There are following arithmetic operators supported by Bourne Shell.

Assume variable a holds 10 and variable b holds 20 then :

Operator	Description	Example
+	Addition - Adds values on either side of the operator	`expr $a + $b` will give 30
–	Subtraction - Subtracts right hand operand from left hand operand	`expr $a - $b` will give -10

*	Multiplication - Multiplies values on either side of the operator	`expr $a * $b` will give 200
/	Division - Divides left hand operand by right hand operand	`expr $b / $a` will give 2
%	Modulus - Divides left hand operand by right hand operand and returns remainder	`expr $b % $a` will give 0
=	Assignment - Assign right operand in left operand	a=$b would assign value of b into a
==	Equality - Compares two numbers, if both are same then returns true.	[$a == $b] would return false.
!=	Not Equality - Compares two numbers, if both are different then returns true.	[$a != $b] would return true.

It is very important to note here that all the conditional expressions would be put inside square braces with one spaces around them, for example [$a == $b] is correct where as [$a==$b] is incorrect.

5.14.2 Relational Operators

Bourne Shell supports following relational operators, which are specific to numeric values. These operators would not work for string values unless their value is numeric.

Assume variable a holds 10 and variable b holds 20 then :

Operator	Description	Example
-eq	Checks if the value of two operands are equal or not, if yes then condition becomes true.	[$a -eq $b] is not true.
-ne	Checks if the value of two operands are equal or if values are not equal then condition becomes true.	[$a -ne $b] is true.
-gt	Checks if the value of left operand is greater than the value of right operand, if yes then condition becomes true.	[$a -gt $b] is not true.
-lt	Checks if the value of left operand is less than the value of right operand, if yes then condition become true.	[$a -lt $b] is true.
-ge	Checks if the value of left operand is greater than or equal to the value of right operand, if yes then condition become true.	[$a -ge $b] is not true.
-le	Checks if the value of left operand is less than or equal to the value of right operand, if yes then condition becomes true.	[$a -le $b] is true.

5.14.3 Boolean Operators

There are following Boolean operators supported by Bourne Shell.

Assume variable a holds 10 and variable b holds 20 then :

Operator	Description	Example
!	This is logical negation. This inverts a true condition into false and vice versa.	[! false] is true.
-o	This is logical OR. If one of the operands is true then condition would be true.	[$a -lt 20 -o $b -gt 100] is true.
-a	This is logical AND. If both the operands are true then condition would be true otherwise it would be false.	[$a -lt 20 -a $b -gt 100] is false.

5.14.4 String Operators

There are following string operators supported by Bourne Shell.

Assume variable a holds "abc" and variable b holds "efg" then :

Operator	Description	Example
=	Checks if the value of two operands are equal or not, if yes then condition becomes true.	[$a = $b] is not true.
!=	Checks if the value of two operands are equal or not, if values are not equal then condition becomes true.	[$a != $b] is true.
-z	Checks if the given string operand size is zero. If it is zero length then it returns true.	[-z $a] is not true.
-n	Checks if the given string operand size is non-zero. If it is non-zero length then it returns true.	[-z $a] is not false.
str	Check if str is not the empty string. If it is empty then it it returns false.	[$a] is not false.

5.14.5 File Test Operators

There are following operators to test various properties associated with a Unix file.

Assume a variable file holds an existing file name "test" whose size is 100 bytes and has read, write and execute permission on :

Operator	Description	Example
-b file	Checks if file is a block special file if yes then condition becomes true.	[-b $file] is false.
-c file	Checks if file is a character special file if yes then condition becomes true.	[-b $file] is false.
-d file	Check if file is a directory if yes then condition becomes true.	[-d $file] is not true.
-f file	Check if file is an ordinary file as opposed to a directory or special file if yes then condition becomes true.	[-f $file] is true.
-g file	Checks if file has its set group ID (SGID) bit set if yes then condition becomes true.	[-g $file] is false.

-k file	Checks if file has its sticky bit set if yes then condition becomes true.	[-k $file] is false.
-p file	Checks if file is a named pipe if yes then condition becomes true.	[-p $file] is false.
-t file	Checks if file descriptor is open and associated with a terminal if yes then condition becomes true.	[-t $file] is false.
-u file	Checks if file has its set user id (SUID) bit set if yes then condition becomes true.	[-u $file] is false.
-r file	Checks if file is readable if yes then condition becomes true.	[-r $file] is true.
-w file	Check if file is writable if yes then condition becomes true.	[-w $file] is true.
-x file	Check if file is execute if yes then condition becomes true.	[-x $file] is true.
-s file	Check if file has size greater than 0 if yes then condition becomes true.	[-s $file] is true.
-e file	Check if file exists. Is true even if file is a directory but exists.	[-e $file] is true.

5.15 CONTROLLING TERMINAL INPUTS

Most UNIX commands take input from the terminal keyboard and send output to the terminal monitor. To redirect output to a file, use the > symbol.

For example,

```
$ls > myfiles
```

Lists the files in your current directory and places them in a file called myfiles. Likewise, you can redirect input with the < symbol. For example,

```
$wc -l < myfiles
```

Feeds the command wc with input from the file myfiles. Although you could obtain the same output by having the filename as an argument, the need for input redirection becomes more apparent in shell programming.

To string the output from one command to the input of the next command, you can use the | (pipe) symbol. For example,

```
$ls -s | sort -nr | pg
```

This lists the files in the current directory with blocksize and then pipes the output to the sort, which sorts the files in numeric descending order and pipes that output to the paging command pg for final display on the terminal's monitor. The pipe command is one of the most useful tools when creating command constructs.

5.15.1 Output Redirection to a File

The > redirection operator sends the output of a command to a file.

For example

```
$ ls -l >otfile
```

In this example shell to save the output of ls –l as the file otfile. If a file with that name already exists in the current directory then it is overwritten its contents are emptied and replaced by the output of the command. If the file otfile does not exist then the shell created it before running the command.

The >> operator appends data to a file without overwriting it. It means if the file already exists then the new data will be added on to the end.

For example

```
$cat arya >> anu
```

The shell redirects the standard ouput from cat and appends it to the file named anu. The result is the contents of arya are added to the end of anu, with out replacing another information in the file. If the file does not exist, it will be created.

5.15.2 Input Redirection from a File

As we know that symbol, >(greater than), to redirect standard ouput. We can use the less than symbol, <, to redirect standard input. The symbol (<) less than tells the shell to interpret the filename that follows it as the standard input to command.

The standard input of a command may be taken from a file instead of the terminal by writing.

For example,

```
$wc <newfile
```

The command wc reads its standard input (in this case redirected from newfile) and prints the number of characters, words and lines found. If only the number of lines is required then

```
$wc -l <file
```

could be used.

5.16 TRAPPING SIGNALS

```
trap signal command
trapabort [off]
```

Arrange for *command* to be executed whenever the signal *signal* is received. *signal* may be specified symbolically (e.g. TERM or XCPU) or numerically.

Trapabort arranges for the TCL error handler to be called whenever a segmentation violation, illegal instruction or bus error occurs. Often the error is mild enough, that processing can continue –it is used in particular with the Object Browser. Specifying trapabort off turns off this behaviour.

5.16.1 UNIX Signals

A *signal* is a message which can be sent to a running process.

Signals can be initiated by programs, users, or administrators.

For example, to the proper method of telling the Internet Daemon (inetd) to re-read its configuration file is to send it a SIGHUP signal.

For example, if the current process ID (PID) of inetd is 4140, we would type:

```
$ kill -SIGHUP 4140
```

Another common use of signals is to stop a running process. To stop the inetd process completely, we would use this command:

```
$ kill 4140
```

By default, the kill command sends the SIGTERM signal. If SIGTERM fails, we can escalate to using the SIGKILL signal to stop the process:

```
$ kill -9 4140
```

Because SIGKILL cannot be handled, stopping a process with SIGKILL is generally considered a bad idea. Using SIGKILL prevents a process from cleaning up after itself and exiting gracefully.

5.16.2 Handling Signals

Each Unix signal has a default set of effects on a Unix program. Programmers can code their applications to respond in customized ways to most signals. These custom pieces of code are called *signal handlers*.

Two signals are unable to be redefined by a signal handler. SIGKILL always stops a process and SIGSTOP always moves a process from the foreground to the background. These two signals cannot be "caught" by a signal handler.

Signal Name	Signal Number	Signal Description
SIGHUP	1	Terminal line hangup
SIGINT	2	Interrupt program
SIGQUIT	3	Quit program
SIGILL	4	Illegal instruction
SIGTRAP	5	Trace trap
SIGABRT	6	Abort
SIGEMT	7	Emulate instruction executed
SIGFPE	8	Floating-point exception
SIGKILL	9	Kill program
SIGBUS	10	Bus error
SIGSEGV	11	Segmentation violation
SIGSYS	12	Bad argument to system call
SIGPIPE	13	Write on a pipe with no one to read it
SIGALRM	14	Real-time timer expired
SIGTERM	15	Software termination signal
SIGURG	16	Urgent condition on I/O channel
SIGSTOP	17	Stop signal not from terminal
SIGTSTP	18	Stop signal from terminal
SIGCONT	19	A stopped process is being continued
SIGCHLD	20	Notification to parent on child stop or exit
SIGTTIN	21	Read on terminal by background process
SIGTTOU	22	Write to terminal by background process
SIGIO	23	I/O possible on a descriptor
SIGXCPU	24	CPU time limit exceeded
SIGXFSZ	25	File-size limit exceeded
SIGVTALRM	26	Virtual timer expired
SIGPROF	27	Profiling timer expired
SIGWINCH	28	Window size changed

SIGINFO	29	Information request
SIGUSR1	30	User-defined signal 1
SIGUSR2	31	User-defined signal 2
SIGTHR	32	Thread interrupt

Shell procedures may use the **trap** command to catch or ignore Unix operating system signals. The form of the trap command is:

```
$trap 'command-list' signal-list
```

Several traps may be in effect at the same time. If multiple signals are received simultaneously, they are serviced in ascending order.

To check what traps are currently set use the trap command. For example:

```
trap
```

5.16.3 Examples of Interrupt Handling

To use single quotes to inhibit command substitution:

```
$!/bin/sh
trap 'echo `pwd` >>$HOME/errdir' 2 3 15
for i in /bin /usr/bin /usr/any/bin
do
cd $i
some series of commands in the directory $i
done
```

The file errdir will contain the name of the directory being worked on when the procedure is interrupted. What happens if the same procedure has double quotes around it?

```
trap "echo `pwd` >errdir" 2 3 15
```

The file errdir will just contain the name of the directory from which the procedure was invoked because the pwd command would be substituted on the first scan by the shell and not when it is invoked in the script.

To remove temporary files when a procedure is interrupted:

```
$!/bin/sh
temp=/tmp/file.$$
trap 'rm $temp; exit' 0 1 2 3 15
ls > $temp
.....
```

If any of the named signals are encountered, the command rm $temp; exit will be executed. The **exit** command is needed to terminate the execution of the whole procedure.

To continue processing commands after a **trap** command:

```
$!/bin/sh
$ read and process commands
dir='pwd'
```

```
for i in *
do
   if test -d $dir/$i
   then
   cd $dir/$i
   while echo ''$i:''
   trap exit 2 # trap ^C
   read x
   do
   trap "2 # ignore interrupts
   eval $x
   done
   fi
 done
```

The shell continues to process commands after a trap command. The entire procedure is terminated if interrupted when waiting for input, but the interrupt is ignored while executing a command. The command list is an explicitly quoted null command and so the signal is ignored by the shell.

5.16.4 trap: Handling command lists

The command list is placed between single quotes, as the command line is scanned twice, once when the shell first encounters the trap command and again when it is being executed.

```
$trap 'command-list' signal-list
```

The single quotes inhibit immediate command and variable substitution but are stripped off after the first scan, so that the commands are processed when the command is actually executed.

If *command-list* is not specified, then the action taken on receipt of any signal in the signal-list is reset to the default system action.

If *command-list* is an explicitly quoted null command (' ' or " "), then the signals in signal-list are ignored by the shell.

The *command-list* is treated like a subroutine call. The commands in the list are executed when the signal is trapped and control is then returned to the place at which it was interrupted.

These numbers are also used for identifying the signals to be trapped.

Signal	Signal Number
Exit	0
ctrld	1
Del	2
ctrl\	3
sure kill	9
kill	15

5.17 ARRAY

Shell variable is capable enough to hold a single value. This type of variables is called scalar variables.

Shell supports a different type of variable called an array variable that can hold multiple values at the same time. Arrays provide a method of grouping a set of variables. Instead of creating a new name for each variable that is required, you can use a single array variable that stores all the other variables.

All the naming rules discussed for Shell Variables would be applicable while naming arrays.

5.17.1 Array Values

The difference between an array variable and a scalar variable can be explained as follows.

Let you are trying to represent the names of various students as a set of variables. Each of the individual variables is a scalar variable as follows :

```
NAME01="Tanmay"
NAME02="Ashish"
NAME03="Shivam"
NAME04="Aryan"
NAME05="Anu"
```

We can use a single array to store all the above-mentioned names. Following is the simplest method of creating an array variable is to assign a value to one of its indices. This is expressed as follows :

```
array_name[index]=value
```

Here array_name is the name of the array, index is the index of the item in the array that you want to set, and value is the value you want to set for that item.

For example :

```
NAME[0]="Tanmay"
NAME[1]="Ashish"
NAME[2]="Shivam"
NAME[3]="Aryan"
NAME[4]="Anu"
```

If you are using ksh shell the here is the syntax of array initialization :

```
set -A array_name value1 value2 ... valuen
```

If you are using bash shell the here is the syntax of array initialization :

```
array_name=(value1 ... valuen)
```

5.17.2 Accessing Array Values

After you have set any array variable, you access it as follows :

```
${array_name[index]}
```

For example :

```
$ /bin/sh
```

```
NAME[0]="Tanmay"
NAME[1]="Ashish"
NAME[2]="Shivam"
NAME[3]="Aryan"
NAME[4]="Anu"

echo "First Index: ${NAME[0]}"
echo "Second Index: ${NAME[1]}"
```

This would produce following result :

```
[root]$./test.sh

First Index: Tanmay
Second Index: Ashish
```

You can access all the items in an array in one of the following ways :

```
${array_name[*]}
${array_name[@]}
```

For example :

```
$/bin/sh

NAME[0]="Tanmay"
NAME[1]="Ashish"
NAME[2]="Shivam"
NAME[3]="Aryan"
NAME[4]="Anu"
echo "First Method: ${NAME[*]}"
echo "Second Method: ${NAME[@]}"
```

This would produce following result :

```
[root]$./test.sh
First Method : Tanmay Ashish Shivam Aryan Anu
Second Method : Tanmay Ashish Shivam Aryan Anu
```

In this given first example, the colors red, green, and blue are assigned to the first three positions of an array named colors :

```
$ colors[0]=RED
$ colors[1]=GREEN
$ colors[2]=BLUE
```

Alternatively, we can perform the same assignments using a single command :

```
$ set -A colors RED GREEN BLUE
```

Adding a dollar sign and an opening brace to the front of the general syntax and a closing brace on the end allows you to access individual array elements :

```
${arrayname[subscript]}
```

Using the array we defined above, let's access (print) each array element one by one :

```
$ print ${colors[0]}
RED
$ print ${colors[1]}
GREEN
$ print ${colors[2]}
BLUE
$
```

If we access an array without specifying a subscript, 0 will be used :

```
$ print ${colors[]}
RED
$
```

The while construct can be used to loop through each position in the array :

```
$ i=0
$ while [ $i -lt 3 ]
> do
> print ${colors[$i]}
> (( i=i+1 ))
> done
RED
GREEN
BLUE
$
```

5.18 COMMAND LINE INTERPRETATION

When we log in, starting a special version of a shell called an interactive shell then we see a shell prompt, usually in the form of a dollar sign ($), a percent sign (%), or a pound sign (#). When we type a line of input at a shell prompt, the shell tries to interpret it. Input to a shell prompt is sometimes called a command line. The basic format of a command line is

```
$command arguments
```

Command is an executable UNIX command, program, utility, or shell program. Arguments are passed to the executable command. Most UNIX utility programs expect arguments to take the following form:

```
$command options filenames
```

For example, in the command line

```
$ ls -l file1 file2
```

There are three arguments to `ls`; the first is an option, and the last two are filenames.

One of the most important things the shell does for the kernel is to eliminate unnecessary information. For a computer, one type of unnecessary information is blank space; therefore, it is important to know what the shell does when it sees blank space. Blank space consists of space characters, horizontal tabs, and new line characters.

Consider this example:

```
$ echo Arya A Arya B Arya C
Arya A Arya B Arya C
```

Here, the shell has interpreted the command line as the echo command with six arguments and has removed the blank space between the arguments.

For example, if we were printing headings for a report and wanted to keep the blank space, we would have to enclose the data in quotation marks, as in

```
$ echo Arya A Arya B Arya C
Arya A Arya B Arya C
```

The single quotation mark prevents the shell from looking inside the quotes. Now the shell interprets this line as the echo command with a single argument, which happens to be a string of characters including blank space.

5.19 RESERVED WORDS

All shell versions have words that have special meaning. In shell programming, words such as `do`, `done`, `for`, and `while` provide loop control—and `if`, `then`, `else`, and `if` provide conditional control. Each shell version has different reserved word pertaining to its specific features.

The reserved words in UNIX are given below:

echo	read	set	unset	readonly	shift
export	if	else	fi	while	do
done	for	until	case	esac	break
continue	exit	return	trap	wait	eval
ulimit	umask				

Total reserved words in UNIX are 26.

☞ SUMMARY

Linux and UNIX systems give us a choice of shell, including sh, chs, ksh, tesh and bash etc. They all provide the essential features of a command interpreter and high-level programming languages, but there are some important differences among them different system provide one or another of these as the default shell, but most systems support the others and allow us to choose the shell us prefer. This chapter give us some information that can help us to decide which shell to use as our own.

We know how to use all the important matching, control functions that the shell provides. We can use shell filename matching, control standard input and output, construct command pipelines, run commands in the background, assign shell variables, use simple command aliases, and configure our chosen shell.

In this chapter, we learned the fundamentals of shell programming, including how to write and execute simple shell scripts, how to include UNIX system command in our scripts, and how to pass arguments to the shell. We also learned more advanced techniques, including flow control with if statementss and for, while loops. We saw how expr can be used to evaluate mathematical expressions.

Shell scripting does have limitations. It is not especially good at string or text manipulation for example awk and sed. The shell is also a programming language that executes shell scripts in the interpretive mode-one line at a time. The interpreter line signifies the sub-shell that runs the script.

The read statement is used with one or more variable to provide input to a scriopt from the keyboard. The command line arguments passed to a script are read into positional parameters (like $1, $2, etc.)

☞ EXERCISE

1. What is a shell? How many types of shell program are typically found on a UNIX system? Name the default shell. (UPTU 2000, 04)
2. Explain the different types of variables in UNIX operating system. What is the meaning of following variables in UNIX operating system?
   (i) PS2, (ii) Home, (iii) Term, (iv) Path (UPTU, 2002)
3. Determine what the built-in shell variables are for your account. Which ones are exported. (UPTU, 2002, 04)
4. What is Shell Variables? Explain in brief.
5. What is Wild cards? Explain in brief.
6. Discuss the utility of sheel metacharacters using suitable example.
7. Explain the usage of following variables :
   (i) $# (ii) $* (iii) $$ (iv) PSI (v) $?
8. Distinguish between Bourne Shell, CShell and Korn Shell.
9. Find the result of the following :

```
For i in xyz
    do
    echo #i
done
```

10. What is shell Script? Why to write shell script?
11. Explain shell arithmatic operators with an example.
12. Write a shell program to reverse the digits of an input number "n".
13. Explain all the decision making statement in details.
14. Write short note on array in UNIX.
15. Write a shell script to find the factorial value of any number entered through keyboard.

❏❏❏

CHAPTER 6

UNIX SYSTEM ADMINSTRATION

6.1 INTRODUCTION

System administrator is concerned with day to day aspects of looking after UNIX system. As we know that UNIX is a multi user operating system so that it requires the environment to be control very carefully and restricted. Under normal circumstance, system administration is an area with which the average user doesn't need to be concerned.

We can also define the UNIX system Administrator as follow:

1. One who administers; one who manages, carries on, or directs the affairs of any establishment or institution.
2. One who has the faculty of managing or organizing.
3. One who executes or performs the official duties of religion, justice, etc.
4. One to whom authority is given to magnetic.

6.1.1 Some Common System Administration Tasks

1. Performing backups
2. Adding and removing users
3. Adding and removing hardware
4. Restoring files from backups that users have accidentally deleted
5. Installing new software
6. Answering users questions
7. Monitoring system activity, disc use and log files
8. Figuring out why a program has stopped working since yesterday, even though the user didn't change anything; honest!
9. Monitoring system security

10. Adding new systems to the network
11. Talking with vendors to determine if a problem is yours or theirs, and getting them to fix it when it *is* their problem
12. Figuring out why the network is so slow.
13. Trying to free up disc space
14. Rebooting the system after a crash.
15. Writing scripts to automate as many of the above tasks as possible

6.2 FILE SYSTEM MOUNTING AND UNMOUNTING

The file system is defined as a tree. Where '/' denoted as a root and /dev, /usr, and the other directories in the root directory are branches, which may have their own branches, such as /usr/local, and so on.

Under UNIX we can think of the filesystem as everything being a file. Thus directories are really nothing more than files containing the names of other files and so on. In addition, the filesystem is used to represent physical devices such as tty lines or even disk and tape units.

Each file on the system has what is called an inode that contains information on the file. UNIX records three file times in the inode, these are referred to as ctime, mtime, and atime. The ctime field refers to the time the inode was last changed, mtime refers to the last modification time of the file, and atime refers to the time the file was last accessed.

The ctime file of the inode is updated whenever the file is written too, protections are changed, or the ownership changed. Usually, ctime is a better indication of file modification than the mtime field. The mtime and atime fields can easily be changed via a system call in C. The ctime field is a little harder to change, although not impossible.

1. The fstab File

During the boot process, file systems listed in /etc/fstab are automatically created (Mounted). The /etc/fstab file contains a list of lines of the following format:

```
device
/mount-point
fstype
options
dumpfreq
passno
```

2. device

A device name which should exist.

3. mount-point

A directory (which should exist), on which to create (mount) the file system.

4. fstype

The file system type to pass to mount(). The default this file system is ufs.

5. options

Either rw for read-write file systems, or for read-only file systems, followed by any other options that may be needed.

6. dumpfreq

This is used by dump() to determine which file systems require dumping. If the field is missing, a value of zero is assumed.

7. passno

This determines the order in which file systems should be checked. File systems that should be skipped should have their passno set to zero. The root file system should have it's passno set to one, and other file systems' passno should be set to values greater than one.

6.2.1 The mount : Mount File System

The mount command is ultimately used to create (mount) file system so that the kernel is aware that the file system is there. Even if the device which contains the specified file system also contains other file system. Therefore any given file system can be moved to a different logical device with no trouble.

The mount command is used like this:

```
$ /etc/mount device mountpoint(Dirname)
    or
$mount [-p | -v]
```

where device is the name of the device on which the file system is to be created (mounted), and dirname is the name of the directory under which that file system is to be created. Directory must already exist. It is created by mkdir command

For example, Suppose that in the file system we just created a floppy disk. We will create a new directory under the root, and mount the file system on the floppy under that directory.

```
$ cd ..
$ mkdir anu
$ /etc/mount /dev/f11 /anu
```

Mount Options

-a Mount all the file systems listed in `/etc/fstab`. Exceptions are those already mounted.

-d Do everything except for the actual system call. This option is useful in conjunction with the -v flag to determine what mount is actually trying to do.

-f Force the mount of an unclean file system (dangerous), or a file system's mount status from read-write to read-only.

-r Mount the file system read-only. This is identical to using the `rdonly` argument

-u Update mount options on the file system.

-w Mount the file system read-write.

-v For display the file name or which we processed.

6.2.2 The umount: Unmount File System

The umount command is used to unmount (demount) a file system or device.

The umount command is used like this:

```
$ /etc/umount device
    Or
$umount -a [ -V ] [ -o specific_options ] [ mount_point... ]
```

where device is the name of the file system which is to be unmounted.

For example, suppose that the file system on the /dev/f11 floppy disk is to be removed.

```
$cd ..
$/etc/umount /dev/f11
```

First of all, we changed our current working directory to the root of the file system. Then we go the umount command as shown.

The umount command takes, as a parameter, one of a mountpoint, a device name, or the -a or -A option.

All forms take `-f` to force unmounting, and `-v` for verbosity. Be warned that `-f` is not generally a good idea. Forcibly unmounting file systems might crash the computer or damage data on the file system.

`-a` and `-A` are used to unmount all mounted file systems, possibly modified by the file system types listed after `-t`. `-A`, however, does not attempt to unmount the root file system.

6.3 SYSTEM BOOTING

To initialize a system from an inactive state, an administrator goes through a bootstrap sequence. The administrator boots the system. Boot procedures vary according to machine types, but the goal is common to all machines to get a copy of the operating system into machine memory and start executing it.

A UNIX system has several different modes of operating system called (system states) and this state is represented by a number or letter, called the run level. The kernel start another process init and default run levels as well as the action to take for each run level are controlled by init.

6.3.1 System State Summary (or Init Level)

0	Shutdown System state
1	Administrative System mode state.
2	Multi user state
3	Network file system state or Full multi-user mode
4	User defined mode (Not defined by system)
5 or S	Single user state
6	Stop and reboot state

6.3.2 Boot Phases

(i) ROM monitor may run some simple hardware diagnostics and finds boot device

(ii) ROM monitor loads/runs bootstrap loader

(iii) Bootstrap loader loads/runs kernel (PID 0)

(iv) Kernel may perform more elaborate diagnostics, then checks root file system

(v) Kernel mounts root file system, then starts init(PID 1)

(vi) Operator may need to switch to multi-user mode

(vii) init runs initialization scripts.

(viii) getty/login produces prompts on terminals and system is ready to go

When our system is up and running, we may decide that we want to change the current state.

For example, you are in single mode and want to change to multi-user mode type the following:

```
$init 2
```

All level 2 multi-user processes will be started.

6.4 SHUTTING DOWN

We can shut down system using the init command, however it is more common to use the shutdown command. Shutdown can be used not only power down the computer, but also go to lower state. The benefit of using shutdown is that it lets us assign a grace period so that users will have some warring before the shutdown actually begins.

The following example of the shutdown command:

```
$shutdown +60 "Shutting down for weekly backups"
$shutdown-g 60 –I 6 "System is shutting down. Please log off carefully"
```

Tells the system to shut down after waiting for a grace period of one minute(-g 60), and to stop the UNIX system and reboot immediately to the level defined by initdefault in the inittab file (-I 6).

Shutdown usually performs the following activities:

(i) Notifies users with wall about the system going shutdown.

(ii) Send signals to all running processes so that they can terminate normally.

(iii) Logs users off and kills remaining processes.

(iv) Writes information about file system status to disk to preserve the integrity of the file system.

(v) Notifies users to reboot or switch off, or moves the system to single-user mode.

System Crashes

(i) Hardware failures

(ii) Power failures/surges

(iii) Environmental problems

(iv) I/O problems (including filled partitions)

(v) Software problems (kernel faults, bugs, programs)

6.5 HANDLING USER ACCOUNT

The term user in UNIX is not mean to be only a person, it can represent an application or a project as well. A group of users performing similar functions may use the same username to use the system. Handling user account UNIX provide four command. These are: **useradd**, **usermod, userdel** and **groupadd**

6.5.1 useradd: Adding new user

There are few ways to add users to your UNIX system. Once is to use the menu interface for your system and follow the prompts. Most UNIX variants use the **useradd** command to identify a new user to the system and allow the new user to access the system.

These are some instruction for maintaining user account:

(i) Determine login name, user ID (UID), group ID (GID)
(ii) Enter data in /etc/passwd file
(iii) Assign password
(iv) Set account parameters (e.g., password aging, account expiration, quotas)
(v) Set file ownerships and protections (use chown, chgrp, chmod commands)
(vi) Add user to other facilities
(vii) Perform other site-specific initialization tasks
(viii) Test new account

The following is an example of how to add a user with user name of *anu*

```
$useradd -m anu
```

This will define the new user *anu* using information from the default user environment. The –m option will create a home directory for the user in */home/anu*

This is a simple method for adding new user in this example apply only one option. But we can use more than one option at a time.

All parameters related to the user have to be provided in the command itself:

```
$useradd -u 143 -g DK -c "MY NAME" -d /home/anu -s /bin/sh -m anu
```

In this procedure creates the user anu with a UID of 143 and group name DK. The home directory is /home/anu, and user will use the Bourne shell. The –m option ensures that the home directory is created if it doesn't already exist.

useradd options

We can include the following information when we add a new user:

Option	Meaning
-u	This sets the user ID(UID).
-o	Use this option with –u to assign a UID that is not unique.
-g	This sets an existing group,s ID or group name.
-d	This sets the home directory of the new users.
-s	This sets the full path of the user's login shell.
-c	Use this to set nay command you want to add to the user's.
-f	this sets the number of the days alogin can be inactive before it declared invalid.

User Password

A new login locked until a password is added for it. You add initial passwords for every regular user.

```
$passwd username
```

A command sequence to do this would look like this:

```
$useradd -m anu
$passwd abc
```

Enter new password for login

New password:

```
$
```

User Profile

All user information except the password encryption is now stored in etc/passwd. The encryptions itself stored in /etc/shadow. There are seven filed in the user profile. These are given below:

1. **User Name:** The name you use to log on to a UNIX system.
2. **UID:** This is user identification number.
3. **Password :** This is password for security.
4. **GID :** The user/s group identification number.
5. **Comment :** User detail such as Name, Address and so on.
6. **Home Directory:** The directory where the user ends up on logging in.
7. **Login shell :** The first program executed after logging in.

6.5.2 userdel & usermod: Removing and modifying users

Users can be removed from the system with userdel. The following command removes user *anu* from the system:

```
$userdel anu
```

This removes all entries connected or containing to anu from /etc/passwd, /etc/group and /etc/shadow. The user's home directory doesn't get deleted in the process and has to be removed separately if required.

Usermod is used for modifying some of the parameters set with useradd. If we want to change the login shell then we can change the login shell by using usermod command. The following command line sets the Korn as the login shell for the user *anu*

```
$usermod -s /bin/ksh anu
```

6.5.3 Adding Group : groupadd

Creating group can be useful in the cases where you want a number of users to have permissions to a particular set of files or directories. If a user is to be placed in a new group, an entry for the group has to be created first in /etc/group.

To add a group called DK to your system, type the following:

```
$groupadd DK
```

This command will add the bane DK to the /etc/group file, and a group assign a group ID number

6.5.4 Deleting Group : groupdel

If any group exists in the UNIX system and no need for longer used, we can delete it this way:

```
$groupdel DK
```

The command will delete the name DK from the /etc/group file.

6.5.5 Modifying the Group: groupmod

If we want to change a group name, we can use groupmod command:

$groupmod –n arya DK

This will change the group's name from DK to arya.

6.6 BACKUP AND RECOVERY

Backup and recovery or restore procedures are critically important. They protect you from losing the valuable data on your computer. Backup is a procedure for coping system data from the permanent storage medium on your computer to another medium. Several different approaches are taken to doing backups and restores.

These are cpio, tar and dump.

6.6.1 CPIO: Copy Input Output

The cpio command accepts its input from standard input and directs its output to standard output. The cpio command copies files to and from a backup devices. As we know it uses standard input to take the list of filenames. The cpio command operates in three modes:

1. cpio-o output mode
2. cpio-i Input mode
3. cpio-p Pass through mode

1. CPIO Command with Output Mode

You give cpio-o a list of files and a destination, cpio then package those files into a single cpio archive file and copies that archive to the destination.

Since cpio use only standard input, you can use is to generate a list of filenames to serve as its input. The –o option creates the archive on the standard output, which you need to redirect to a device file.

```
$ls | cpio -ov > /dev/rdsk/f0dt
Array
Calander
Different
Convert
Abcd.sh
278 Blocks
Total size of the archive
```

The –v(verbose) option displays each filename on the terminal while it's being copied. Cpio needs as input a list of the files, and if this list is available in a file, redirect can be used too:

```
$cpio -o >/dev/rdsk/f0dt < first
```

(i) Incremental Backups

Find can also produce a file list, so any files that satisfy its selection criteria can also be backed up. We will frequently need to use find cpio in combination to back up selected files. Those that have been modifies in the last two days:

```
$find. - types f -mtime -2 -print | cpio -ovB >/dev/rdsk/f0dt
```

After find command dot shows the files are back up with their relative pathnames. If it is a /, find will use absolute pathnames. The –B option sets the block size to 5120 bytes for input, which is 10 times the default size.

(ii) Multivolume Backups

When the created archive in the backup device is larger than the capacity of the device, cpio prompts for inserting a new diskette into drive:

```
$find . -types f -print | cpio -ocB >/dev/rdsk/fodt
```

If we want to go on, type device/folename when ready

```
/dev/fd0
3672 blocks
```

Copy in/out archive program, Must be given a list of file names to archive. Use find to do this:

```
find . -print | cpio -ocBV > /dev/tape
```

reloading a archive from tape

```
cpio -icdv < /dev/tape
```

2. CPIO command with Input Mode

We give cpio –I a cpio archive files via standard input. cpio then splits the archive back into the separate file s and directories and replaces them on the system. Selected files or a complete archive can be restored with the –i option. Like this:

```
$cpio -iv < /dev/rdsk/fodt
Array
Calander
Different
Convert
Abcd.sh
278 Blocks
```

When restoring subdirectories, cpio assumes that the subdirectory structures are also mentioned on the hard disk. cpio also accepts wild cord pattern, so multiple file fitting the pattern can be restored.

```
$cpio -i "*.sh" < /dev/rdsk/f0dt
```

3. CPIO Command with Pass-through Mode

The cpio –p command form works like the output mode, except that instead of copying the files to an archive, it copies each file individually to another directory in the UNIX system file system tree.

The following example uses cpio –p to pass copies of files from one point in the directory structure to another point.

```
$cd /home/mcn
$find . -dept -print | cpio -pmvd /mnt/mcnbackup
```

Exact copies of all files and directories below /home/mcn are passed to the /mnt/mcnbackup directory.

Displaying the Archive

The –t option displays the contents of the device without restoring the files. The –t option must combined with the –i option:

```
$cpio -itv </dev/rdsk/f0dt
```

100143	anu	101	Jan	14	10:20:35	2008 anu.dk
100143	anu	111	Jan	14	10:20:35	fnumber.pl
100143	anu	123	Jan	14	10:20:36	dec_bin.pl
100143	anu	243	Jan	14	10:20:36	home_get.pl

The files are displayed in a format resembling the listing. This format shows the octal representation of the file type (10) and the permissions and permissions (0755) as well as modification time of the file.

In cpio command we can use the following options:

Option	Meaning
-d	Creates directories.
-c	Writes header information in ASCII.
-r	Renames files in interactive manner.
-t	Lists files in archive.
-u	Overwrites newer file with older version.
-v	Verbose option.
-m	Retains original file modification time.
-c size	Sets input output block size to size bytes.
-f exp	Copies all files except those in exp.

6.6.2 The tar Command: the tape Archive Program

The tape archive program, easy to use and transportable. Advantage is that tar is supported everywhere and also useful for copying directories. The tar command has been in existence since before the emergence of cpio. The tar command versatile command with certain exclusive features not found in cpio:

(i) It does not use standard input to obtain its file list. tar accepts the file and directory names as arguments.

(ii) It copies one or more entire directory trees that are it operates recursively by default.

(iii) It can append to an archive without overwriting the entire archive.

1. Backing up files with –c option

The –c option is used to copy files to the back up device. The –v option shows the progress of backup:

```
$tar -cvf /dev/rdsk/f0dt /home/SQL/*.sql
```

The command backs up all SQL scripts with their absolute pathnames.

2. Multivolume Backup

For multivolume diskette backups, tar uses a special option (-k), followed by the volume size in KB.

```
$tar -cvfkb /dev/rdsk/f0dt 1440 18 index
```

tar estimates that two 1440 KB the argument to –k diskettes will be required. After the first volume is use full, tar prompts for a new volume. At the time of restoration, the same option has to be used.

3. Restoring files with –x option

Files are created with the –x option. When no file or directory name is specified, it restores all files from the backup device. The following command restores the files just back up:

```
$tar -xvfb /dev/rdsk/f0dt 18
```

Selective extraction is also possible by providing one or more filenames or directory. Unlike cpio when files are extracted the modification times of the files also remain unchanged. This can be overridden by the –m option to reflect the system time at the time of extraction.

4. Displaying the Archive with –t option

Just like as cpio command, the –t option display the content of the device in a long format similar to the listing:

```
$tar -tvf /dev/rdsk/fodt
```

rwxr_xr_x101/50 452 Jan 15 09:35 2008 ./anu.dk

The files have been backed up with relative pathnames. Each filename is preceded by ./. We can also use many more options is tar command. Which are given below:

Option	Meaning
-c	Creates a new archive.
-x	Extracts files from archive.
-t	Lists contents of archive.
-r	Appends files at the end of archive.
-u	Like –r option, but only new files.
-w	Conforms from user about action to be taken.
-m	Changes modification time of file to time extraction.

These are the options in the tar command.

6.6.3 The dump Command

The dump (or rdump) reads the raw file system and copies the data blocks out to tape. dump is program that is generally used by most groups. One feature of dump is that dump (via rdump) can work across a network and dump a file system to a remote machine's tape drive. Dump can not be used to back up just a specific directory. Thus for backup of a directory tree tar or cpio is better suited.

Dump uses level numbers to determine whether a full or partial dump should be done. The levels can be between 0 and 9.

The dump command has **many** options. A sample dump command would look like: $dump 0usdf 6000 54000 /dev/tape /users.

where

0 - represent the dump level (0 through 9)

u - signifies to update the /etc/dumpdates file

s - is the size of the tape in feet
d - represents the density of the tape
f - represents the device to use for the dump
6000 - is the size parameter in feet
54000 - is the density parameter
/dev/tape - is the device to use
/users - is the filesystem to dump

6.7 SECURITY

The UNIX system was designed so that users could easily access their resources and share information with other users. Security is also an important feature of the UNIX system. UNIX has always included features to protect it from unauthorized users and to protect user's resources, without impending authorized users. The UNIX has following the security. These are given below:

6.7.1 Setting User ID (SUID) Permission

When we execute program we create a process. Four identifiers are assigned to this process upon its creation. These are its real (User Identification) uid, real (Group Identification) gid effective uid, and effective gid.

Unless the set user ID (suid) permission and / or the set group ID (sgid) permission of an executable file are set. The process created is assigned ud uid and gid as its real and effective uid and real and effective gid, respectively. In this case, the process has exactly the same permissions that we do. For instance, for the process to execute a program, we must have execute permission for the containing this program.

When the suid permission of an executable file is set, a process created from the program has its effective uid set to that of the owner of the file, instead of us own uid. This means that the file access privileges of the process are determined by the permission of the owner of the file. For instance, if the suid permission is set , a process can create a file when the owner of the file has execute permission and write permission for the directory where file will be created.

We can use chmod to set the suid permission of a file

```
$chmod u+s DK
```

Sets the suid permission of DK. DK is a file.

We can also use chmod to remove the suid permission of a file:

```
$chmod u-s DK
```

Removes the suid permission from DK

Similarly we can set the Group ID (GID) permissions.

1. Uses of Suid Permission

The suid permission is used in several important user programs that need to read or write files owned by root. When we run the passwd command to change us password, we have same permissions as root.

2. Problems with SUID

SUID programs have a place on the system, in fact it would be difficult, if not impossible, to run Unix without SUID programs. For example, the file /bin/passwd is a SUID program owned by root. This makes sense, you want to restrict access to changing passwords in some way and

SUID is a good way to do that. That said, SUID programs are the most frequent way that security is compromised. Problems with SUID programs fall into the following two categories.

3. Carelessness

The root user leaves an unattended root window and someone uses that window to create a backdoor into the system.

4. Poor Software Design.

Often programs are written SUID that can be written without being SUID root. Also, many programmers don't fully understand the consequences of the code they write and leave unintended back doors into the system.

6.7.2 Passwd: Changing Password

The command passwd is an important security feature of the UNIX system. Any user can change the old password by using passwd command.

We can change the password like this:

```
$passwd
```

Changing password

Enter the new password (Minimum 5 & Maximum 8 characters)

New password: ******

Re –enter password: ******

Password changes

6.7.3 The Sticky Bit

A process runs by loading the executable file from the system. This means gathering all the blocks of the file which may be scattered throughout the disk and pushing them to memory. The overhead for reading a file from the file system are quite high and to reduce the time of reading, a mode is used which is called the sticky bit. The sticky bit is extremely useful for implementing group projects.

Using chmod, we can set the bit on a directory by using a as the additional bit:

```
$chmod 1775 anu
$ls -l anu
drwxrwxr-t    2    Anita        abc1024 Jan 14 10:30 anu
```

To let a group of users work on a set of files without infirming on security, we will have to do this:

1. Create a command group for these users in /etc/geoup.
2. Create separate user accounts for them but specify the same home directory.
3. Make sure the home directory. Use chown to surrender ownership to root.
4. Make the directories group-writable and their sticky bits with chmod 1775.

6.7.4 Chmod

The chmod command (change mode) sets the file permission for a set of files. The command supports two options -R and -f. The -R option is used to recursively move down through the file system and select files matching the filelist. The -f option is used to keep errors from being reported back (when used in shell scripts). Since an octal number requires three bits each bit of the number corresponds to a file permission (read, write, or execute). Thus using the octal number 7 which has a binary value of 111 which implies read, write, and execute access.

- 4000 - Setuid on execution
- 2000 - setgid on execution
- 1000 - set sticky bit
- 0400 - read by owner
- 0200 - write by owner
- 0100 - execute by owner
- 0040 - read by group
- 0020 - wr
- 0010 - execute by group
- 0004 - read by others
- 0002 - write by others
- 0001 - execute by others

6.7.5 Encrypting Files

To encrypt an SCCII file, such as a text message, using the public key of the intended recipient of the file, use a command of the form:

```
$pgp -e file userid
```

For example, to send the anu.dk to arya, use the command

```
$pgp -e anu.dk arya
```

This will produce a file anu.dk, which is an encrypted version of anu.dk, encrypted using arya's public key. We should also convert the encrypted file to ASCII. This can be done automatically using the –a option. For example,

```
$pgp -ea anu.dk arya
```

Encrypt the file anu.dk using arya's public key and convert the file into ASCII.

PGP also provides the –t option, used to ensure that text messages sent via mail to different types of system.

We should use:

```
$pgp -eat msg.txt arya
```

6.8 CREATING FILES

We can create a unix file by using one of the process :

1. **cat command**
2. **touch command**
3. **using vi editor**

1. cat command

The cat is of the most important command of UNIX. By using this command we can create file, display content of the file and concatenate of more than two files.

For example

```
$ cat >fruits
```

```
apple
orange
grapes
```

(Press control+d)

If we want display the content of the file, then we can also use cat command, like as :

```
$ cat fruits
apple
orange
grapes
```

2. touch command

We can also create Unix file by using touch command.

For example, by using touch command we can create a empty file such as :

```
$ touch < file_name>
(Press Ctl +d)
```

3. using vi editor

To start using vi, at the Unix prompt type vi followed by a file name. If we are creating a new file, type in the name wish to give to the new file.

```
$vi <filename>
```

Then Press the Enter key. We will see a screen similar to the one below which shows blank lines with tildes and the name and status of the file.

```
~
~
"myfile" [New file]
```

6.9 STORAGE OF FILES

Like other common operating systems information on our UNIX systems is stored in files. Besides files, UNIX allows the creation of directories, in which other files and directories can be placed.

Something like :

```
            File System
          /   <-- root directory (a slash)
        /   \
     home   ...
       /  \
  ugrad course
       /     \
   ...     csxxxx
            / | \
        ... ... ☺ username <-- your home directory
```

So that those that have accounts for this course (except CS majors) will have an account in the location:

/home/course/csXXX/username

where XXX is the course number

/home/ugrad/username

This is called the path of your home directory, the directory you are placed in when you log in.

A path just represents the location of a file or directory in the file hierarchy. For example, the first path above says that under the root (/) directory, there is a home directory, under it, a course directory, under it, a csXXX directory, and under that is where your username directory is.

Since this path starts at the root (/), it is called an absolute (or full) path.

6.10 DISK RELATED COMMANDS

6.10.1 df : For find the free Disk Space.

The df stands for disk free. This command is used to show the amount of free space on one or more filesystems.

syntax

```
$df [option]-filename
$df
/ (/dev/root ): 1269 blocks 27856 i-nodes
$df /
```

Displays the free disk space in the / filesystem.

```
$df -kvi
```

A more verbose display showing the free disk space and free i-node space of all filesystems, measured in kilobytes.

Options

Show information about the filesystem on which each FILE resides, or all filesystems by default.

- **-a** : all - include filesystems having 0 blocks
- **-B** : block-size=SIZE use SIZE-byte blocks
- **-h** : human-readable - print sizes in human readable format (e.g., 1K 234M 2G)
- **-H** : si - likewise, but use powers of 1000 not 1024
- **-I** : inodes - list inode information instead of block usage
- **-k** like - block-size=1K
- **-l** : local - limit listing to local filesystems

- **-P** : portability - use the POSIX output format
- **-t** : type=TYPE - limit listing to filesystems of type TYPE
- **-T** : print type - print filesystem type
- **-x** : exclude-type - limit listing to filesystems not of type TYPE
- **-v** : ignored

6.10.2 du : Used Disk Space

The du stands for disk usage. This command is used to show the amount of disk space consumed by one or more files or directories (or directory trees).

```
syntax
    $du [option] - file or directory name
    $du
    122     ./backup
    425     ./fa/backup
    2281    ./fa
    5       ./check
    18      ./dbf
```

We can also find out the usages of subdirectory like as

```
$du /dev
```

Options

- **-a** : All
- **-B** : Block size use size byte blocks
- **-b** : Bytes- print size in bytes
- **-c** : Total - produce a grand total
- **-D** : Deference args- deference FILEs that are symbolic links
- **-h** : Human readable- print sizes in human readable format
- **-H** : si- likewise, but use powers of 1000 not 1024
- **-k** : like block size = 1k
- **-l** : Count links - count sizes many times if hard linked
- **-L** : Deference - deference all symbolic links
- **-S** : Separate dirs - do not include size of subdirectories
- **-s** : Summarize - display only total for each argument
- **-x** : only-file-system - skip directories on different filesystem

6.10.3 dfspace : Fee Disk Space int MB

We can think that our system hard disk if it has 27857 free inodes or 12970 blocks. But don't know the disk space in the term of megabyte or percentage (%). The dfspace does exactly that. It reports the free disk space in term of meghabytes and percentage of total disk space.

```
$dfspace
```

Dfspace: not found

The dfspace command present in/etc directory.

```
$/etc/dfspace
```

: Disk space: 6.30 MB of 135.74 MB available (4.59%)

Total Disk Space: Disk space: 6.30 MB of 135.74 MB available (4.59%)

6.10.4 ulimit : User Limit

The ulimit stand for user limit. This command show the user space

```
$ulimit
2097151
```

This implies that the user can not create a file whose size is bigger that 2097151 bytes.

6.11 USER QUATAS

```
quota - display disk usage and limits
```

Syntax

```
quota [ -F format-name ] [ -guvs | q ]
quota [ -F format-name ] [ -uvs | q ] user
quota [ -F format-name ] [ -gvs | q ] group
```

Quota displays users' disk usage and limits. By default only the user quotas are printed.

Quota reports the quotas of the entire file systems listed in /etc/mtab. For file systems that are NFS-mounted a call to the rpc.rquotad on the server machine is performed to get the information.

OPTIONS

-F format-name

Show quota for specified format. Possible format names are: vfsold (version 1 quota), vfsv0 (version 2 quota), rpc (quota over NFS), xfs (quota on XFS filesystem)

-g

Print group quotas for the group of which the user is a member. The optional

-u

flag is equivalent to the default.

-v

will display quotas on file systems where no storage is allocated.

-s

flag will make quota try to choose units for showing limits, used space and used inodes.

-q

Print a terser message, containing only information on file systems where usage is over quota.

Specifying both -g and -u displays both the user quotas and the group quotas (for the user).

6.12 USER ACCOUNTS

UNIX system must have a login name or username defined to the system. This login name identifies the user. When a new user is added to UNIX systems administrator assigns it a unique login name, which is associated with a number called UID or user identification number. This UID is the systems way of identifying the user. Typically, a new user is also assigned to a group; group is collection of users in one department or working on one project identified with one group name. Each group is identified with its GID number or group identification number.

For example,

$ls -la command will produce following result. in /usr/ssb directory for tanmay.

```
drwxrwxrwx   2 ssb   users   1024 Nov  3 12:50 dusg
-rw-rw-rw-   1 ssb   users   348 Oct 30 09:46 dxi12
drwxrwxrwx   3 ssb   users   1024 Oct 21 10:38 gnu
drwxrwxrwx   2 ssb   users   1024 Sep 18 11:17 logs
-rwxr-xr-x   1 ssb   users   1726 Oct 23 12:11 mon12.ksh
```

It tells us that first dusg is a directory owned by ssb in users group with 1024 bytes created on Nov 3 at 12:50 PM.

So together UID and GID determine what kind of access rights a user has to files and directories.

These are procedures to add a new user to system.

(i) Get user's name and create a login id based on name, decide which group and which other groups' user need.

(ii) Enter this data into /etc/passwd file and /etc/group file.

(iii) Assign a password to this account, also set password aging, account expiration date, resource limits and privileges.

(iv) Create a home directory for user, preferably on the same disk where other home directories exist.

(v) chown and cghrp to give new user ownership of his home directory.

(vi) Set umask for new files that are created, set disk quota, mail and printing system for new user.

(vii) Test this account.

Then at next prompt assign the new password.

(i) -f Force the user to change her password at next login.

(ii) -s List password aging date for the specified user.

(iii) -a Used with -s to list data for all users.

(iv) -l Lock the account so the user can't log in. (-u unlocks it).

(v) -d delete the password for this user.

To create a home directory for this user use mkdir and chown commands.

```
cd /usr/
$mkdir tanmay
$chown tanmay tanmay
```

Since this user gets ksh when logs in we have to give him a default .profile file, if a user logs in with csh then .login file is used. These files are helpful for takes such as

(i) Setting the path search, so that programs can be run which are not in home directory.

(i) file protection with umask value.

(iii) Definition of terminal type and initialization of terminal.

(iv) Other customization functions.

☞ SUMMARY

- The administrator can change the attributes of any file, kill any process and change any user's password. He/She can use every type of service of the UNIX system.
- The administrator is responsible for startup and shutdown of the system. The init process takes the system to multi-user mode, single user mode or any other mode. The shutdown program is used to power down the system.
- The init daemon process maintains the system at one run level. The multi-user mode (2) is used by the administration for checking the file system or installing peripherals.
- Init read all the instructions from /etc/inittab, which contains scripts related to file system checking, mounting, starting and stopping of daemon, and monitoring the system.
- To administrator, it is important to have a basic understanding of the major file system and media on which they are stored.
- A sound understanding of directory structure and backing up and restoring data is also essential.

☞ EXERCISE

1. What are the main jobs that a system administration has to carry out?
2. What do you mean by mounting and unmounting? Explain with suitable example.
3. Explain the important option each of TAR and CPIO.
4. Write short note on backup management.
5. Explain the various security features available in UNIX operating system.
6. What is file system? Discuss the creation mounting and unmounting file system in UNIX.
7. Discuss using suitable examples the backup and restoration of files using cpio. How is it different from tar?
8. What is the various security features offered in UNIX? Explain any two in detail.

❏❏❏

CHAPTER 7

UNIX FILTERS AND EXPRESSIONS

7.1 INTRODUCTION

A filter is a UNIX command that does some manipulation of the text of a file. Four of the most powerful and popular UNIX filters are the grep, sed, awk and perl commands. Four of these commands are extremely powerful and complex. A filter is a program that gets most of its data from its standard input and writes its main results to its standard output. UNIX filters are often used as elements of pipelines. The pipe operator (|) on a command line signifies that the main output of the command to the left is passed as main input to the command on the right.

A filter is a small and specialized program in UNIX operating systems that transforms plain text data in some meaningful way and that can be used together with other filters and pipes to form a series of operations that produces highly specific results.

As is generally the case with command line programs in UNIX operating systems, filters read data from standard input and writes to standard output. Standard input is the source of data for a program, and by default it is text typed in at the keyboard. However, it can be redirected to come from a file or from the output of another program. Standard output is the destination of output from a program, and by default it the display screen. This means that if the output of a command is not redirected to a file or another device such as a printer or piped to another filter for further processing, it will be sent to the monitor where it will be displayed.

Numerous filters are included on Unix systems, a few of which are awk, cat, comm, csplit, cut, diff, expand, fold, grep, head, join, less, more, paste, sed, sort, spell, tail, tr, unexpand, uniq and wc etc.

7.2 GENERALIZED REGULAR EXPRESSION PARSER (GREP)

grep is a command line text search utility originally written for Unix. The name is taken from the first letters in global / regular expression / print, a series of instructions in text editors such as ed. A acronym of the unusual name also exists in the form of Generalized Regular Expression Parser. The grep command searches files or standard input globally for lines matching a given regular expression, and prints them to the program's standard output.

Syntax

The syntax for the grep command is :

```
$grep [options] pattern [files]
```

7.2.1. Searching a pattern

grep searches out these patterns globally and prints out the lines containing them.

The most searching form of grep is :

```
$grep [pattern] [filename]
```

Here grep searches the named file for the the pattern and prints out the lines containing the pattern.

For example :

```
$grep lovely tanmay.txt
```

In this case, grep prints all lines containing lovely from the file tanmay.txt, regardless of word boundaries; therefore lines containing dearlovely or lovely are also printed. The grep command is case sensitive by default, so this example's output does not include lines containing Lovely (with a capital L) unless they also contain lovely.

To search all .txt files in a directory for lovely in a shell that supports globbing, use an asterisk (*) in place of the file name :

```
$ grep lovely *.txt
```

Regular expressions can be used to match more complicated queries. The following prints all lines in the file that begin with the letter a, followed by any one character, then the letters ple.

```
$grep ^a.ple fruitlist.txt
```

As noted above, the term grep derives from a usage in ed and related text editors. Before grep existed as a separate command, the same effect might have been achieved by doing :

```
$ed fruitlist.txt
g/^a.ple/p
q
```

Where the second line is the command given to ed to print the relevant lines, and the third line is the command to exit from ed.

Like most UNIX commands, grep accepts options in the form of command-line arguments, to change many of its behaviors.

7.2.2. Input for grep

If file name is not given after the pattern, then grep takes its input from the standards input. The reason for providing for standard input is that this allows grep to receive input from a pipe(|).

For example, you wanted to know if arya was logged in and didn't wish to read through the whole who output yourself, you could type.

```
$who | grep arya
```

7.2.3. Options in grep

The grep command allows you to search one file or multiple files for lines that contain a pattern. Exit status is 0 if matches were found, 1 if no matches were found, and 2 if errors occurred.

The main options for grep :

options	Meaning
-b	Display the block number at the beginning of each line.
-c	Display the number of matched lines.
-h	Display the matched lines, but do not display the filenames.
-i	Ignore case sensitivity.
-l	Display the filenames, but do not display the matched lines.
-n	Display the matched lines and their line numbers.
-s	Silent mode.
-v	Display all lines that do NOT match.
-w	Match whole word.

For example,

```
$grep -i apple fruitlist.txt
```

This prints all lines containing apple regardless of capitalization. The -i argument tells grep to be case insensitive, or to ignore case.

To print all lines containing apple as a word (pineapple and apples will not match) :

```
$grep -w apple fruitlist.txt
```

But if fruitlist.txt contains apple word followed by hyphen (-) character, it will also get matched.

```
$cat fruitlist.txt
apple
apples
pineapple
apple-
apple-fruit
fruit-apple
$grep -w apple fruitlist.txt
apple
apple-
apple-fruit
fruit-apple
```

The -v (lower-case v) prints all lines that do NOT contain apple in this example.

```
$grep -v apple fruitlist.txt
```

banana

pear

peach

orange

7.3 THE STREAM EDITOR (SED)

UNIX editors like vi and emacs, sed is a non-interactive stream-oriented editor. This non-interactive feature allows you to use sed to automate editing sequences if desired. The name sed is an abbreviation for stream editor, and the utility derives many of its commands from the ed line-editor.

A Stream Editor is used to perform basic transformations on text read from a file or a pipe. The result is sent to standard output. The syntax for the sed command has no output file specification, but results can be saved to a file using output redirection. The editor does not modify the original input. You can modify streams of text on the fly. So the work cycle of sed is for each line :

1. read a entire line from stdin in its pattern buffer
2. modify the pattern buffer according to the supplied commands
3. print the pattern buffer to stdout

sed uses Basic Regular Expressions. Basic Regular Expressions are very similar to other types of Regular Expressions; in fact many users won't see any difference at all.

Syntax

```
$sed [options] program [inputfile]
```

You do not need to interact with the editor while it is running; that is why sed is sometimes called a batch editor. This feature allows use of editing commands in scripts, greatly easing repetitive editing tasks. When facing replacement of text in a large number of files, sed is a great help.

The sed program can perform text pattern substitutions and deletions using regular expressions, like the ones used with the grep command.

The editing commands are similar to the ones used in the vi editor :

Command	Result
a\	Append text below current line.
c\	Change text in the current line with new text.
d	Delete text.
i\	Insert text above current line.
p	Print text.
r	Read a file.
S	Search and replace text.
W	Write to a file.

Apart from editing commands, you can give options to sed. An overview is in the table given below :

Option	Effect
-e SCRIPT	Add the commands in SCRIPT to the set of commands to be run while processing the input.
-f	Add the commands contained in the file SCRIPT-FILE to the set of commands to be run while processing the input.
-n	Silent mode.
-V	Print version information and exit.

The following simple program consists of one command only d. The command d tells sed to delete the pattern buffer.

```
$ sed -e 'd' /etc/hosts
```

When you launch this script nothing happens. Remember the general workflow of sed: read a line into the pattern buffer, elaborate the line according to the script and then print the line to stdout. In this case the pattern buffer is deleted by the script, and no output was generated.

Another command is p. It tells sed to print the pattern buffer.

```
$ sed -e 'p' /etc/hosts
```

The effect of this script is to print the line twice. Remember the operation mode of sed :

1. read a line from stdin
2. print the line (because of the p command)
3. print the pattern buffer to stdout

sed commands can be issued on the command line using the following syntax :

```
$sed [-e] 'command1' [-e command2...] [file]
```

Notice that more than one sed command can be performed by prefacing each command with the -e option. This option is not required if only one command is used. Specifying a file is optional because sed can take its input from either files or stdin (standard input).

Similar to a UNIX shell script, multiple sed commands may also be stored in a script file. The -f option is used on the command line to access the commands in the script :

```
$sed -f script [file]
```

The s command replaces the first occurrence of UNIX on each line in the file with UNIX. Adding g to the end of the s command will instruct sed to perform a global substitution:

```
$ sed 's/UNIX/unix/g' LFL
```

Sed command in UNIX is commonly used for processing of files. Sed stands for Stream Editor which parses text files and used for making textual transformations to a file. The command specified to Sed, is applied on the file line by line.

Example

To replace all matching occurrences of some text to another.

```
$sed -e 's/olddata/newdata/g' inputFileName > outputFileName
```

Here, g is global which replaces ALL occurrences.

Sed command reads from a standard input and places it into the pattern space. It performs various editing commands on that pattern space in a sequential manner. Later the pattern space is written onto the STDOUT.

7.3.1. Addresses

We don't always want to apply a command work on every single line. Sometimes we want to apply a command to a single line, or a block of line. sed provides a mechanism to address one line or more lines. This mechanism is called an address.

An address is one of the following :

Options	Meaning
n	selects line number n
$	selects the last line
/re/	selects the lines matching the Regular Expression re
\crec	selects the lines matching the Regular Expression re. The c may be any character
first~step	GNU extension! Selects every step'th line starting with line first
addr1,addr2	Address range: selects all input lines which match the inclusive range of lines starting from the first address and continuing to the second address
addr!	select lines that do not match addr

Examples

The command = prints the current line number. A substitute program for wc -l might be :

```
$sed -n -e '$ ='
```

Both examples that follow emulate the UNIX program head :

```
$ sed -n -e `1,10p`
$ sed -e `10q`
```

The first example uses the address 1-10 to select the lines to print. The second example uses the implicit print command at each cycle to provide the output. When the address '10' matches, sed will be terminated.

7.3.2. Substitution Command

Eliminate comments

```
$ sed -e `s/#.*//`/etc/inetd
```

Eliminate comments and empty lines

```
$ sed -e 1s/#.*//1/etc/inetd
```

7.3.3. Comments

The character # is a command which cannot have any address. This is useful if the sed-program is stored in a file. The whole program can be executed with comments

```
$sed -f programfile < inputdata
```

7.3.4. Space Balls

- The patterns are manipulated in the pattern space
- The hold space can store multiple lines, separated by newline.
- There are commands to fill/empty the hold space
- There aren't any commands to work directly on the hold space

Options	Meaning
D	Delete text in the pattern space up to the first newline
N	Add a newline to the pattern space, then append the next line of input to the pattern space
P	Print out the portion of the pattern space up to the first newline
h	Replace the contents of the hold space with the contents of the pattern space
H	Append a newline to the contents of the hold space, and then append the contents of the pattern space to that of the hold space
g	Replace the contents of the pattern space with the contents of the hold space
G	Append a newline to the contents of the pattern space, and then append the contents of the hold space to that of the pattern space
x	Exchange the contents of the hold and pattern spaces

Example

Print the first line as last

```
$sed -n -e '1h;1!p;${g;p;}'
```

h: hold space <- pattern space

g: pattern space <- hold space

7.4 AWK

We are ready covered grep and sed in this section 7.2 and 7.3. In this section we will discusses AWK. AWK is an excellent filter and report writer. AWK is an excellent tool for processing these rows and columns. AWK also has string manipulation functions, so AWK can search for particular strings and modify the output. AWK also has associative arrays, which are incredible useful. Associative arrays can make a complex problem a trivial exercise.

Many UNIX utilities have strange names. AWK is one of those utilities. It is not an abbreviation for awkward. In fact, it is an elegant and simple language. awk is a pattern-matching, programmable and processing tool available in UNIX. It works equally well with text and numbers.

7.4.1. awk Filtering

awk is not just a command, this is a programming language also. In other words, awk utility is a pattern scanning and processing language. It searches one or more files to see if they contain lines that match specified patterns and then perform associated actions, such as writing the line to the standard output or incrementing a counter each time it finds a match.

Syntax

```
$awk option 'selection {action}' file(s)
```

Here, selection filters input and select lines for the action component to act upon. The selection is enclosed within single quotes and the action within the curly braces. Both the selection and action forms an awk program.

For example,

Example 1

```
$ awk '/manager/ { print }' emp.lst
```

Output

9001	Tanmay	Manager	IT
2355	Rakesh	Manager	Production
5683	Rohit	Manager	Sales

In the above example, /manager/ is the selection which selects lines that are processed in the action section i.e. {print}. Since the print statement is used without any field specifiers, it prints the whole line.

Since printing is the default action of awk, any one of the following three forms can be used :

```
$awk '/manager/ ' emp.lst
$awk '/manager/ { print }' emp.lst
$awk '/manager/ { print $0}' emp.lst
0 specifies complete line.
```

(i) Pattern Matching

Awk uses regular expression in sed style for pattern matching.

For example,

Example 2

```
$awk -F "|" ' /r [ao]*/' emp.lst
```

Output

2355	Rakesh	Manager	Production
5683	Rohit	Manager	Sales

7.4.2. Splitting a line into fields

Awk uses special parameter, $0, to indicate entire line. It also uses $1, $2, $3 to identify fields. These special parameters have to be specified in single quotes so that they will not be interpreted by the shell.

awk uses contiguous sequence of spaces and tabs as a single delimiter.

For example,

Example 3

```
$awk -F "|" '/production/ { print $2, $3, $4 }' emp.lst
```

Output

```
Rakesh | Manager | Productions
```

Rahul	|	Accountant	|	Productions
Shivam	|	Clerk	|	Productions

In the above example-3, comma (,) is used to delimit field specifications to ensure that each field is separated from the other by a space so that the program produces a readable output.

We can also specify the number of lines we want using the built-in variable NR as illustrated in the following example :

Example 4

```
$awk -F "|" 'NR==2, NR==4 { print NR, $2, $3, $4 }' emp.lst
```

Output

2	Rakesh	Manager	Productions
3	Rahul	Accountant	Productions
4	Shivam	Clerk	Productions

7.4.3. printf: Formatting Output

The printf statement can be used with the awk to format the output. Awk accepts most of the formats used by the printf function of C.

For example,

Example 5

```
$awk -F "|" '/[kK]u?[ar]/ { printf "%3d %-20s %-12s \n", NR, $2, $3}'
>emp.lst
```

Output

4	D Kumar	Manager
8	S Kumar	Accountant
4	A Kumar	Clerk

Here, the name and designation have been printed in spaces 20 and 12 characters wide respectively.

Redirecting Standard Output

The print and printf statements can be separately redirected with the > and | symbols. Any command or a filename that follows these redirection symbols should be enclosed within double quotes.

For example,

Exampl 6 :

```
use of |
printf "%3d %-20s %-12s \n", NR, $2, $3 | "sort"
```

Example 7 :

```
use of >
printf "%3d %-20s %-12s \n", NR, $2, $3 > "newlist"
```

7.4.4. Variables and Expressions

Variables and expressions can be used with awk as used with any programming language. Here, expression consists of strings, numbers and variables combined by operators.

For example,

Example 8 :

```
(x+2)*y, x-15, x/y, etc..,
```

A variable is an identifier that references a value. To define a variable, you only have to name it and assign it a value. The name can only contain letters, digits, and underscores, and may not start with a digit. Case distinctions in variable names are important: Name and name are two different variables. awk allows the use of user-defined variables without declaring them i.e. variables are deemed to be declared when they are used for the first time itself.

Example 9 :

```
X= "4"
x= "3"
Print X
Print x
```

Strings in awk are enclosed within double quotes and can contain any character. Awk strings can include escape sequence, octal values and even hex values. Octal values are preceded by \ and hex values by \x. Strings that do not consist of numbers have a numeric value of 0.

Example 10 :

```
z = "Dear"
print z
prints Dear
```

Example 11 :

```
y = "\t\t Dear \7"
print y
```

prints two tabs followed by the string Dear and sounds a beep.

String concatenation can also be performed. Awk does not provide any operator for this; however strings can be concatenated by simply placing them side-by-side.

Example 12 :

```
z = "Hi" "Arya"
print z
prints Hi Arya
```

Example 13 :

```
p = "Tanmay" ; q= "Arya"
print p q
prints Tanmay Arya
```

Example 14 :

```
x = "Tanmay"
y = "Arya"
print x "&" y
prints Tanmay & Arya
```

A numeric and string value can also be concatenated.

Example 15 :

```
l = "8" ; m = "2" ; n = "Dear"
Print l m
```

prints 82 by converting m to string.

```
Print l - m
```

prints 6 by converting l as number.

```
Print m + n
```

prints 2 by converting n to numeric 0.

Expressions also have true and false values associated with them. A nonempty string or any positive number has true value.

Example 16 :

```
if(c)
```

This is true if c is a nonempty string or positive number.

7.4.5. The Comparison Operators

awk also provides the comparison operators like >, <, >=, <= ,==, !=, etc.

Example 16 :

```
$ awk -F "|" '$3 == "manager" || $3 == "chairman" {
> printf "%-20s %-12s %d\n", $2, $3, $5}' emp.lst
```

Output

```
T Arya      chairman   15000
J.P.Gola    manager    9000
Ashu        manager    8750
Anu         manager    8500
```

The above command looks for two strings only in the third filed ($3). The second attempted only if (||) the first match fails.

Note: awk uses the || and && logical operators as in C and UNIX shell.

Example 17 :

```
$ awk -F "|" $3 != "manager" && $3 != "chairman" {
> printf "%-20s %-12s %d\n", $2, $3, $5}' emp.lst
```

Output

S kumar	Accountant	7000
A Kumar	Clerk	6000
Rahul	Accountant	7000
Rakesh	Clerk	6000

The above example illustrates the use of != and && operators. Here all the employee records other than that of manager and chairman are displayed.

(a) ~ and !~ : The Regular Expression Operators

In awk, special characters, called regular expression operators or metacharacters, can be used with regular expression to increase the power and versatility of regular expressions. To restrict a match to a specific field, two regular expression operators ~ (matches) and !~ (does not match).

Example 18 :

```
$2 ~ /[cC]ho[wu]dh?ury / || $2 ~ /sa[xk]s ?ena /Matches second field
```

Example 19 :

$2 !~ /manager | chairman / Neither manager nor chairman

Note:	The operators ~ and !~ work only with field specifiers like $1, $2, etc.,.

For instance, to locate g.m.s the following command does not display the expected output, because the word g.m. is embedded in d.g.m or c.g.m.

```
$ awk -F "|" '$3 ~ /g.m./ {printf "…..
```

prints fields including g.m like g.m, d.g.m and c.g.m

To avoid such unexpected output, awk provides two operators ^ and $ that indicates the beginning and end of the filed respectively. So the above command should be modified as follows :

```
$ awk -F "|" '$3 ~ /^g.m./ {printf "...
```

prints fields including g.m only and not d.g.m or c.g.m

The following table depicts the comparison and regular expression matching operators.

Table 7.1 : Comparison and regular expression matching operators

Operator	Significance
<	Less than
<=	Less than or equal to
==	Equal to
!=	Not equal to
>=	Greater than or equal to
>	Greater than
~	Matches a regular expression
!~	Doesn't matches a regular expression

(b) Number Comparison

Awk has the ability to handle numbers (integer and floating type). Relational test or comparisons can also be performed on them.

Example 20:

```
$ awk -F "|" '$5 > 7500 {
> printf "%-20s %-12s %d\n", $2, $3, $5}' emp.lst
```

Output

T Arya	chairman	15000
J.P. Gola	manager	9000
Rohit	manager	8750
Rakesh	manager	8500

In the above example, the details of employees getting salary greater than 7500 are displayed.

Regular expressions can also be combined with numeric comparison.

Example 21 :

```
$ awk -F "|" '$5 > 7500 || $6 ~/1980$/' {
> printf "%-20s %-12s %d\n", $2, $3, $5, $6}' emp.lst
```

Output :

T Arya	chairman	15000	30/12/1982
J.P. Gola	manager	9000	01/01/1980
Rohit	manager	8750	10/05/1985
Rakesh	manager	8500	20/05/1988
Rahul	Accountant	6000	01/10/1980
Anil	Clerk	5000	20/05/1980

In the above example, the details of employees getting salary greater than 7500 or whose year of birth is 1980 are displayed.

(c) Number Processing

Numeric computations can be performed in awk using the arithmetic operators like +, -, /, *, % (modulus). One of the main feature of awk w.r.t. number processing is that it can handle even decimal numbers, which is not possible in shell.

Example 22 :

```
$ awk -F "|" '$3' == "manager" {
> printf "%-20s %-12s %d\n", $2, $3, $5, $5*0.4}' emp.lst
```

Output

J P Gola	manager	9000	3600
Rohit	manager	8750	3500
Rakesh	manager	8500	3250

In the above example, DA is calculated as 40% of basic pay.

7.4.6. Variables

Awk allows the user to use variables of there choice. You can now print a serial number, using the variable kount, and apply it those directors drawing a salary exceeding 6700 :

```
$ awk -F"|" '$3 == "director" && $6 > 6700 {
```

- kount =kount+1
- printf " %3f %20s %-12s %d\n", kount,$2,$3,$6 }' empn.lst

The initial value of kount was 0 (by default). That's why the first line is correctly assigned the number 1. awk also accepts the C- style incrementing forms :

```
Kount ++
Kount +=2
Printf "%3d\n", ++kount
```

(a) The -f Option: Storing Awk Programs In A File

You should holds large awk programs in separate file and provide them with the .awk extension for easier identification. Let's first store the previous program in the file empawk.awk :

```
$ cat empawk.awk
```

Observe that this time we haven't used quotes to enclose the awk program. You can now use awk with the -f filename option to obtain the same output :

```
$Awk -F"|" -f empawk.awk empn.lst
```

(b) The Begin And End Sections

Awk statements are usully applied to all lines selected by the address, and if there are no addresses, then they are applied to every line of input. But, if you have to print something before processing the first line, for example, a heading, then the BEGIN section can be used gainfully. Similarly, the end section useful in printing some totals after processing is over.

The BEGIN and END sections are optional and take the form

```
BEGIN {action}
END {action}
```

These two sections, when present, are delimited by the body of the awk program. You can use them to print a suitable heading at the beginning and the average salary at the end. Store this program, in a separate file **empawk2.awk**

Like the shell, awk also uses the # for providing comments. The BEGIN section prints a suitable heading , offset by two tabs (\t\t), while the END section prints the average pay (tot/kount) for the selected lines. To execute this program, use the -f option :

```
$awk -F"|" -f empawk2.awk empn.lst
```

Like all filters, awk reads standard input when the filename is omitted. We can make awk behave like a simple scripting language by doing all work in the BEGIN section. This is how you perform floating point arithmetic :

```
$ awk 'BEGIN {printf "%f\n", 22/7 }'
3.142857
```

This is something that you can't do with expr. Depending on the version of the awk the prompt may be or may not be returned, which means that awk may still be reading standard input. Use [ctrl-d] to return the prompt.

(c) Built-in Variables

Awk has several built-in variables. They are all assigned automatically, though it is also possible for a user to reassign some of them. You have already used NR, which signifies the record number of the current line. We'll now have a brief look at some of the other variable.

1. ***The FS Variable :*** as stated elsewhere, awk ues a contiguous string of spaces as the default field delimeter. FS redefines this field separator, which in the sample database happens to be the |. When used at all, it must occur in the BEGIN section so that the body of the program knows its value before it starts processing :

   ```
   BEGIN {FS="|"}
   ```

   This is an alternative to the -F option which does the same thing.

2. ***The OFS Variable :*** when you used the print statement with comma-separated arguments, each argument was separated from the other by a space. This is awk's default output field separator, and can reassigned using the variable OFS in the BEGIN section :

   ```
   BEGIN { OFS="~" }
   ```

   When you reassign this variable with a ~ (tilde), awk will use this character for delimiting the print arguments. This is a useful variable for creating lines with delimited fields.

3. ***The NF variable*** : NF comes in quite handy for cleaning up a database of lines that don't contain the right number of fields. By using it on a file, say emp.lst, you can locate those lines not having 6 fields, and which have crept in due to faulty data entry :

   ```
   $awk 'BEGIN { FS = "|" }
   ```

   - NF !=6 {
   - Print "Record No ", NR, "has ", "fields"}' empx.lst

4. ***The FILENAME Variable :*** FILENAME stores the name of the current file being processed. Like grep and sed, awk can also handle multiple filenames in the command line. By default, awk doesn't print the filename, but you can instruct it to do so :

   ```
   '$6<4000 {print FILENAME, $0 }'
   ```

   With FILENAME, you can device logic that does different things depending on the file that is processed.

7.4.7. ARRAYS

An array is also a variable except that this variable can store a set of values or elements. Each element is accessed by a subscript called the index. Awk arrays are different from the ones used in other programming languages in many respects :

(i) They are not formally defined. An array is considered declared the moment it is used.

(ii) Array elements are initialized to zero or an empty string unless initialized explicitly.

(iii) Arrays expand automatically.

(iv) The index can be virtually any thing: it can even be a string.

(a) Associative arrays

Even though we used integers as subscripts in the tot [ ] array, awk doesn't treat array indexes as integers. Awk arrays are associative, where information is held as key-value pairs. The index is the key that is saved internally as a string. When we set an array element using mon[1]="mon", awk converts the number 1 to a string. There's no specified order in which the array elements are stored. As the following example suggests, the index "1" is different from "01" :

```
$ awk 'BEGIN {
•    direction ["N"] = "North" ; direction ["S"] ;
•    direction ["E"] = "East" ; direction ["W"] = "West" ;
•    printf("N is %s and W is %s \n", direction["N"], direction ["W"]);
•    Mon[1] = "Jan"; mon["1"] = "january" ; mon["01"] = "JAN" ;
•    Printf("mon is %s\n", mon[1]);
•    Printf("mon[01] is also %s\n",mon[01]);
•    Printf("mon[\"1\"] is also %s \n", mon["1"]);
•    Printf("But mon[\"01\"] is %s\n", mon["01"]);
}
```

There are two important things to be learned from this output. First, the setting with index "1" overwrites the setng made with index 1. Accessing an array element with subscript 1 and 01 actually locates the element with subscript "1". Also note that mon["1"] is different from mon["01"].

(b) ENVIRON[] : The Environment Array:

You may sometimes need to know the name of the user running the program or the home directing awk maintains the associative array, ENVIRON[], to store all environment variables. This POSIX requirement is met by recent version of awk including nawk and gawk. This is how we access the shell variable , HOME and PATH, from inside an awk program :

```
$nawk 'BEGIN {
>print "HOME" "=" ENVIRON["HOME"]
>print "PATH" "=" ENVIRON["PATH"]
>}'
```

7.4.8. FUNCTIONS

Awk has several built in functions, performing both arithmetic and string operations. The arguments are passed to a function in C-style, delimited by commas and enclosed by a matched pair of parentheses. Even though awk allows use of functions with and without parentheses (like printf and printf()), POSIX discourages use of functions without parentheses.

Some of these functions take a variable number of arguments, and one (length) uses no arguments as a variant form. The functions are adequately explained here so you can confidently use them in perl which often uses identical syntaxes.

There are two arithmetic functions which a programmer will except awk to offer. int calculates the integral portion of a number (without rounding off),while sqrt calculates square root of a number. awk also has some of the common string handling function you can hope to find in any language.

There are :

(a) length : it determines the length of its arguments, and if no argument is present, the enire line is assumed to be the argument. You can use length (without any argument) to locate lines whose length exceeds 1024 characters :

```
awk -F"|" 'length > 1024' empn.lst
```

you can use length with a field as well. The following program selects those people who have short names :

```
awk -F"|" 'length ($2) < 11' empn.lst
```

(b) index(s1, s2) : it determines the position of a string s2within a larger string s1. This function is especially useful in validating single character fields. If a field takes the values a, b, c, d or e you can use this function n to find out whether this single character field can be located within a string abcde :

```
x = index ("abcde", "b")
```

This returns the value 2.

(c) substr (stg, m, n) : it extracts a substring from a string stg. m represents the starting point of extraction and n indicates the number of characters to be extracted. Because string values can also be used for computation, the returned string from this function can be used to select those born between 1946 and 1951 :

```
$awk -F"|" 'substr($5, 7, 2) > 45 && substr($5, 7, 2) < 52' empn.lst
    2365|Tanmay Arya|director|personel|11/05/47|7800|2365
    3564|Anu Arya|executive|personnel|06/07/47|7500|2365
    4290|J P Gola|executive|production|07/09/50|6000|9876
    9876|Ashish Gola|director|production|12/03/50|7000|9876
```

You can never get this output with either sed or grep because regular expressions can never match the numbers between 46 and 51. Note that awk does indeed posses a mechanism of identifying the type of expression from its context. It identified the date field string for using substr and then converted it to a number for making a numeric comparison.

(d) split(stg, arr, ch) : it breaks up a string stg on the delimiter ch and stores the fields in an array arr[]. Here's how yo can convert the date field to the format YYYYMMDD :

```
$awk -F "|" '{split($5, ar, "/"); print "19"ar[3]ar[2]ar[1]}' empn.lst
    19521212
    19501203
    19431904
    ...........
```

You can also do it with sed, but this method is superior because it explicitly picks up the fifth field, whereas sed would transform the only date field that it finds.

(e) system : you may want to print the system date at the begging of the report. For running a UNIX command within a awk, you will have to use the system function. Here are two examples :

For example :

```
BEGIN {
    system("tput clear")  Clears the screen
    system("date")     Executes the UNIX date command
    }
```

7.4.9. CONTROL FLOW

1. The If Statement

Awk has practically all the features of a modern programming language. It has conditional structures (the if statement) and loops (while or for). They all execute a body of statements depending on the success or failure of the control command. This is simply a condition that is specified in the first line of the construct.

Function	Description
int(x)	returns the integer value of x
sqrt(x)	returns the square root of x
length	returns the complete length of line
length(x)	returns length of x
substr(stg, m, n)	returns portion of string of length n, starting from position m in string stg.
index(1s, s2)	returns position of string s2 in string s1
splicit(stg, arr, ch)	splicit string stg into array arr using ch as delimiter, returns number of fields.
System("cmd")	runs UNIX command cmd and returns its exit status

The if statement can be used when the && and || are found to be inadequate for certain tasks. Its behavior is well known to all programmers. The statement here takes the form :

```
If (condition is true) {
     Statement
} else {
Statement
}
```

Like in C, none of the control flow constructs need to use curly braces if there's only one statement to be executed. But when there are multiple actions take, the statement must be enclosed within a pair of curly braces. Moreover, the control command must be enclosed in parentheses.

An alternative form of this logic places the condition inside the action component rather than the selection criteria. But this form requires the if statement :

```
Awk -F "|" '{ if ($6 > 7500) printf .........
```

It can be used with the comparison operators and the special symbols ~ and !~ to match a regular expression. When used in combination with the logical operators || and &&, awk programming becomes quite easy and powerful. Some of the earlier pattern matching expressions are rephrased in the following, this time in the form used by if :

```
if ( NR > = 3 && NR <= 6 )
if ( $3 == "director" || $3 == "chairman" )
if ( $3 ~ /^g.m/ )
if ( $2 !~ / [aA]gg?[ar]+wal/ )
```

```
if ( $2 ~[cC]ho[wu]dh?ury|sa[xk]s?ena/ )
```

To illustrate the use of the optional else statement, let's assume that the dearness allowance is 25% of basic pay when the latter is less than 600, and 1000 otherwise. The if-else structure that implants this logic looks like this :

```
If ( $6 < 6000 )
      da = 0.25*$6
else
      da = 1000
```

You can even replace the above if construct with a compact conditional structure :

```
$6 < 6000 ? da = 0.25*$6 : da = 1000
```

This is the form that C and perl use to implement the logic of simple if-else construct. The ? and : act as separator of the two actions.

When you have more than one statement to be executed, they must be bounded by a pair of curly braces (as in C). For example, if the factors determining the hra and da are in turn dependent on the basic pay itself, then you need to use terminators :

```
If ( $6 < 6000 ) {
      hra = 0.50*$6
      da = 0.25*$6
}else {
      hra = 0.40*$6
      da = 1000
}
```

2. LOOPING WITH for :

awk supports two loops - for and while. They both execute the loop body as long as the control command returns a true value. For has two forms. The easier one resembles its C counterpart. A simple example illustrates the first form :

```
for (k=0; k<=9; k+=2)
```

This form also consists of three components; the first component initializes the value of k, the second checks the condition with every iteration, while the third sets the increment used for every iteration. for is useful for centering text, and the following examples uses awk with echo in a pipeline to do that :

```
$echo "
>Income statement\nfor\nthe month of august, 2002\nDepartment : Sales"
|
>awk ' { for (k=1 ; k < (55 -length($0)) /2 ; k++)
>printf "%s"," "
>printf $0}
Income statement
      for
```

```
the month of August, 2002
Department : Sales
```

The loop here uses the first printf statement to print the required number of spaces (page width assumed to be 55). The line is then printed with the second printf statement, which falls outside the loop. This is useful routine which can be used to center some titles that normally appear at the beginning of a report.

Using for with an Associative Array

The second form of the for loop exploits the associative feature of awk's arrays. This form is also seen in perl but not in the commonly used languages like C and java. The loop selects each indexof an array :

```
for ( k in array )
      commamds
```

Here, k is the subscript of the array arr. Because k can also be a string, we can use this loop to print all environment variables. We simply have to pick up each subscript of the ENVIRON array :

```
$ nawk 'BIGIN {
>for ( key in ENVIRON )
>print key "=" ENVIRON [key]
>}'
LOGNAME=tanmay
MAIL=/var/mail/tanmay
PATH=/usr/bin::/usr/local/bin::/usr/ccs/bin
TERM=xterm
HOME=/home/tanmay
SHELL=/bin/bash
```

Because the index is actually a string, we can use any field as index. We can even use elements of the array counters. Using our sample databases, we can display the count of the employees, grouped according to the disgnation (the third field). You can use the string value of $3 as the subscript of the array kount[] :

```
$awk -F'|' '{ kount[$3]++ }
>END { for ( desig in kount)
>print desig, kount[desig] }' empn.lst
g.m          4
chairman     1
executive    2
director     4
manager      2
d.g.m        2
```

The program here analyzes the databases to break up of the employees, grouped on their designation. The array kount[] takes as its subscript non-numeric values g.m., chairman, executive, etc.. **for** is invoked in the END section to print the subscript (desig) and the number of occurrence of the subscript (kount[desig]). Note that you don't need to sort the input file to print the report!

3. LOOPING WITH while

The **while** loop has a similar role to play; it repeatedly iterates the loop until the command succeeds. For example, the previous **for** loop used for centering text can be easily replaced with a **while** construct :

```
k=0
while (k < (55 - length($0))/2) {
     printf "%s"," "
     k++
}
print $0
```

The loop here prints a space and increments the value of k with every iteration. The condition (k < (55 - length($0))/2) is tested at the beginning of every iteration, and the loop body only if the test succeeds. In this way, entire line is filled with a string spacesbefore the text is printed with **print $0**.

Not that the **legth** function has been used with an argument ($0). This **awk** understands to be the entire line. Since length, in the absence of arguments, uses the entire line anyway, $0 can be omitted. Similarly, **print $0** may also be replaced by simply **print**.

Programs

1. awk script to delete duplicate

```
lines in a file.
BEGIN { i=1;}
{
     flag=1;
          for(j=1; j<i && flag ; j++ )
     {
          if( x[j] == $0 )
          flag=0;
     }
if(flag)
     {
          x[i]=$0;
          printf "%s \n",x[i];
          i++;
     }
}
```

```
Run1 (Execute):
[root@localhost shellprgms]$ cat >for7.txt
hello
world
world
hello
this
is
this
```

Output

```
[root@localhost shellprgms]$ awk -F "|" -f 11.awk for7.txt
hello
world
this
is
```

2. awk script to print the transpose of a matrix

```
BEGIN{
    system("tput clear")
    count =0
    }
{
        split($0,a);
    for(j=1;j<=NF;j++)
{ count = count+1
    arr[count] =a[j]
}
    K=NF
}
END{
printf("Transpose\n");
for(j=1;j<=K;j++)
{
    for(i=j; i<=count; i=i+K)
{
            printf("%s\t", arr[i]);
        }
```

```
            printf("\n");
    }
}
```

Run1 (Execute) :

```
[root@localhost shellprgms]$ qwk -f p8.awk
    2   3
    5   6
Transpose
    2   5
    3   6
```

3. Awk script that folds long line into 40 columns. Thus any line that exceeds 40 Characters must be broken after 40th and is to be continued with the residue. The inputs to be supplied through a text file created by the user.

```
BEGIN{
start=1; }
{ len=length;
for(i=$0; length(i)>40; len-=40)
{
print substr(i,1,40) "\\"
i=substr(i,41,len);
}
print i; }
```

Run1 (Execute) :

```
[root@localhost shellprgms]$ awk -F "|" -f 15.awk sample.txt
aaaaaaaaaaaaaaaaaaaaaaaaaaaaaaaaaaaaaaaa\
aaaaaaaaaaaaaaaaaaaaaaaaaaaaaaaaaaaaaaaa\
aaaaaaaaaaaa
aaaaaaaaaaaaaa
aaaaaaaaaaaaaaaaaaaaaaaaaaaaaaaaaaaaaaaa\
aaaaaaaaa
```

Output

```
[root@localhost shellprgms]$ cat sample.txt
aaaaaaaaaaaaaaaaaaaaaaaaaaaaaaaaaaaaaaaaaaaaaaaaaaaaaaaaaaaaaaaaaaaaaaaaaaaaaaaaaaaaaaaaaaaaa
aaaaaaaaaaaa
aaaaaaaaaaaaaa
aaaaaaaaaaaaaaaaaaaaaaaaaaaaaaaaaaaaaaaaaaaaaaaaa
```

4. This is an awk program to provide extra spaces at the end of the line so that the line length is maintained as 127.

```
awk ' { y=127 - length($0)
      printf "%s", $0
      if(y > 0)
      for(i=0;i<y;i++)
      printf "%s", " "
      printf "\n"
      }' foo
```

5. A file contains a fixed number of fields in the form of space-delimited numbers. This is an awk program to print the lines as well as total of its rows.

```
awk '{ split($0,a)
      for (i=1;i<=NF;i++) {
      row[NR]+=a[$i]
      }

      printf "%s", $0
      printf "%d\n", row[NR]
      } ' foo
```

7.5 PERL

Perl is a programming language optimized for scanning arbitrary text files, extracting information from those text files, and printing reports based on that information. It's also a good language for many system management tasks. The language is intended to be practical rather than beautiful.

If you have used shell scripts or awk or similar UNIX utilities for various purposes, you will find that you can normally use Perl for those and many other purposes and the code tend to be more compact. And if you haven't used such utilities but have started thinking you might have need for them.

Syntax

```
$perl [options] [programfile] [argument]
```

Options

S	Enables rudimentary option parsing for options on the command line after the script name but before any file name arguments.
-T	Forces taint checks to be turned on so you can test them.
-U	Allows perl to do unsafe operations.
-v	Prints the version and patch level configuration of your perl executable.
-V[:conigvar]	Prints summary of the major perl configuration values and the current value of @INC.

-c	Checks the syntax of the script and then exits without executing it.
-w	Prints warnings about variable names that are mentioned only once, and scalar variables that are used before being set.
-d	Runs the script under the perl debugger
-D	Sets debugging flags. To turn on debugging flags, you can either specify a number which is the total of the numeric values of the desired flags. The available flags are : 1 p Tokenizing and parsing 2 s Stack snapshots 4 l Lable stack processing 8 t Trace execution 16 o Operator node construction 32 c String/numeric conversions 64 p Print processor command for -P 128 m Memory allocation 256 f Format processing 512 r Regular expression parsing 1024 x Syntax tree dump 2048 u Tainting checks 4096 L Memory leaks (not supported with later versions) 8192 H Hash dump -- usurps values () 16384 X Scratchpad allocation 32768 D Cleaning up

-p	assumes the following loop around your script.
	while(<>){
	# your script goes here
	} continue {
	print or die "-p destination: $!\n";
	}

-n	Assumes the following loop around your script.
	while (<>) {
	# your script goes here
	}

-a	Turns on autosplit mode when used with a -n or -p.
-F pattern	Specifies a pattern expression to split on
-l octal	Enables automatic line-ending processing.
-0 octal	Specifies the input record separator ($/) as an octal number. If there are no digits, the null character is the separator. Other options may precede or follow the digits.

-I dir	Prepends directory to the search path for modules (@INC)
-m module	Executes use module (); before executing the script.
-M 'module'	executes use module ; before executing the script.
-P	Runs your script through the preprocessor script cppstdin before compilation by perl.
-S	Uses the PATH environment variable to search for the script.
-x dir	Tells perl that the script is embedded in a message.
-i extension	specifies that files processed by the <> construct are to be edited in-place.
-e command	Specifies a line of script.
programfile	Name of the perl program

7.5.1. Writing Perl Program

We will develop first Perl program will show how to display a line of text on your monitor. This is not a very elaborate program but here we focus on the mechanism of writing and running Perl programs :

- First, you create a text file to hold the Perl program.
- Then you run or execute the Perl program file.

(a) Creating the Program

A Perl program consists of an ordinary text file containing a series of Perl statements. Statements are written in what looks like a combination of C, UNIX shell script, and English.

Perl code can be quite free-flowing. The broad syntactic rules governing where a statement starts and ends are :

(i) Leading spaces on a line are ignored. You can start a Perl statement anywhere you want: at the beginning of the line, indented for clarity (recommended) or even right-justified.

(ii) Statements are terminated with a semicolon.

(iii) Spaces, tabs, and blank lines outside of strings are irrelevant-one space is as good as a hundred. That means you can split statements over several lines for clarity. A string is basically a series of characters enclosed in quotes.

(iv) Anything after a hash sign (#) is ignored except in strings. Use this fact to interleave your code with useful comments.

Here's Our first uninspired Perl statement hello1.pl :

```
print("Hello World\n");
```

No prizes for guessing what happens when Perl runs this code-it prints out Hello World. If the "\n" doesn't look familiar, don't worry-it simply means that Perl should print a newline character after the text, or in other words, go to the start of the next line, exactly like C.

Printing more text is a matter of either stringing together statements like this, or giving multiple arguments to the print() function :

```
print("Hello World,\n");
print("I'm alive\n");
```

So here's a small finished example hello2.pl, complete with the invocation line at the top and a few comments :

```
$!/usr/local/bin/perl -w
print("Hello World,\n");
print("I'm alive\n");
```

(b) Comments in Your Program

It is very important to place comments into your Perl programs. Comments will enable you to figure out the intent behind the mechanics of your program.

Comments are placed inside a program file using the # character. Everything after the # is ignored. For example comment.pl :

```
# This whole line is ignored.
print("Perl is easy.\n"); # Here's a half-line comment.
```

7.5.2. Data Types

Perl has a number of fundamental data types. The most commonly used and discussed are scalars, arrays, hashes, filehandles, and subroutines :

1. **scalars :** A scalar is a single value; it may be a number, a string, or a reference.
2. **srrays :** An array is an ordered collection of scalars.
3. **hashes :** A hash, or associative array, is a map from strings to scalars; the strings are called keys, and the scalars are called values.
4. **filehandles :** A file handle is a map to a file, device, or pipe that is open for reading, writing, or both.
5. **subroutines :** A subroutine is a piece of code that may be passed arguments, be executed, and return data.

1. Scalar values

String values must be enclosed by quotes. Enclosing a string in double quotes allows the values of variables whose names appear in the string to automatically replace the variable name in the string. Enclosing a string in single quotes prevents variable interpolation. If $name is "Tanmay", print("My name is $name") will print "My name is Tanmay", but print('My name is $name') will print "My name is $name".

To include a double quotation mark in a string, precede it with a backslash or enclose the string in single quotes. To include a single quotation mark, precede it with a backslash or enclose the string in double quotes. Strings can also be quoted with the q and qq quote-like operators. 'this' is identical to q(this) and "$this" is identical to qq($this).

Finally, multiline strings can be defined using here documents :

```
$multilined_string = <<EOF;
This is my multilined string
note that I am terminating it with the word "EOF".
EOF
```

Numbers (numeric constants) do not require quotation. Perl will convert numbers into strings and vice versa depending on the context in which they are used. When strings are converted into numbers, trailing non-numeric parts of the strings are discarded. If no leading part of a string is numeric, the string will be converted to the number 0. In the following example, the strings $n and

$m are treated as numbers. This code prints the number '5'. The values of the variables remain the same. Note that in Perl, + is always the numeric addition operator. The string concatenation operator is the period.

```
$n = '3 apples';
$m = '2 oranges';
print $n + $m;
```

Functions are provided for the rounding of fractional values to integer values: int chops off the fractional part, rounding towards zero; POSIX::ceil and POSIX::floor round always up and always down, respectively. The number-to-string conversion of printf "%f" or sprintf "%f" round ot even, use bankers' rounding.

Perl also has a boolean context that it uses in evaluating conditional statements. The following values all evaluate as false in Perl :

```
$false = 0; # the number zero
$false = 0.0; # the number zero as a float
$false = 0b0; # the number zero in binary
$false = 0x0; # the number zero in hexadecimal
$false = '0'; # the string zero
$false = ""; # the empty string
$false = undef; # the return value from undef
$false = 2-3+1 # computes to 0 which is converted to "0" so it is false
```

All other values evaluated to true. This includes the odd self-describing literal string of "0 but true", which in fact is 0 as a number, but true when used as a boolean. All non-numeric strings also have this property, but this particular string is truncated by Perl without a numeric warning. A less explicit but more conceptually portable version of this string is '0E0' or '0e0', which does not rely on characters being evaluated as 0, because '0E0' is literally zero times ten to the power zero.

Evaluated boolean expressions are also scalar values. The documentation does not promise which particular value of true or false is returned. Many boolean operators return 1 for true and the empty-string for false. The defined() function determines whether a variable has any value set. In the above examples, defined($false) is true for every value except undef.

If either 1 or 0 are specifically needed, an explicit conversion can be done:

```
my $real_result = $boolean_result ? 1 : 0;
```

2. Array values

An array value is specified by listing its elements, separated by commas, enclosed by parentheses.

```
@scores = (32, 45, 16, 5);
```

The qw() quote-like operator allows the definition of a list of strings without typing of quotes and commas. Almost any delimiter can be used instead of parentheses. The following lines are equivalent :

```
@names = ('Tanu', 'Anu', 'Dev-Anu');
@names = qw(Tanu Anu Dev-Anu);
```

The split function returns a list of strings, which are split from a string expression using a delimiter string or regular expression.

```
@scores = split(',',' '32,45,16,5');
```

Individual elements of a list are accessed by providing a numerical index in square brackets. The scalar sigil must be used. Sublists (array slices) can also be specified, using a range or list of numeric indices in brackets. The array sigil is used in this case.

For example, $month[3] is "March", and @month[4..6] is ("April", "May", "June").

3. Hash values

A hash may be initialized from a list of key/value pairs. If the keys are separated from the values with the => operator, rather than a comma, they may be unquoted (barewords). The following lines are equivalent :

```
$favorite = ('Tanmay', "red", 'Arya', "blue");
$favorite = (Tanmay => 'red', Arya => 'blue');
```

Individual values in a hash are accessed by providing the corresponding key, in curly braces. The $ sigil identifies the accessed element as a scalar. For example, $favorite{Tanmay} equals 'red'. A hash can also be initialized by setting its values individually :

```
$favorite{Tanmay} = 'red';
$favorite{Arya} = 'blue';
$favorite{Anu} = 'green';
```

Multiple elements may be accessed using the @ sigil instead. For example, @favorite{'Tanmay', 'arya'} equals ('red', 'blue').

4. Typeglob values

A typeglob value is a symbol table entry. The main use of typeglobs is creating symbol table aliases. For example :

```
*PI = \3.141592653; # creating constant scalar $PI
*this = *that; # creating aliases for all data types 'this' to all data types 'that'
```

5. Array functions

The number of elements in an array can be determined either by evaluating the array in scalar context or with the help of the $# sigil. The latter gives the index of the last element in the array, not the number of elements. The expressions scalar(@array) and ($#array + 1) are equivalent.

6. Hash functions

There are a few functions that operate on entire hashes. The keys function takes a hash and returns the list of its keys. Similarly, the values function returns a hash's values.

```
# Every call to each returns the next key/value pair.
# All values will be eventually returned, but their order
# cannot be predicted.
while (($name, $address) = each %addressbook) {
    print "$name lives at $address\n";
}
# Similar to the above, but sorted alphabetically
```

```
foreach my $next_name (sort keys %addressbook) {
    print "$next_name lives at $addressbook{$next_name}\n";
}
```

7. Subroutines

Subroutines are defined with the sub keyword and are invoked simply by naming them. If the subroutine in question has not yet been declared, invocation requires either parentheses after the function name or an ampersand (&) before it. But using & without parentheses will also implicitly pass the arguments of the current subroutine to the one called, and using & with parentheses will bypass prototypes.

```
# Calling a subroutine
# Parentheses are required here if the subroutine is defined later in the code
foo();
&foo; # (this also works, but has other consequences regarding arguments passed to the subroutine)
# Defining a subroutine
sub foo { ... }
foo; # Here parentheses are not required
```

A list of arguments may be provided after the subroutine name.

7.5.3. Variables

Perl, like all other computer languages, uses variables to keep track of the usage of computer memory. Every time you need to store a new piece of information, you assign it to a variable.

You have already seen how Perl uses numbers, strings, and arrays. Now, you will see how to use variables to hold this information. Perl has three types of variables :

1. Scalar

denoted by a $ symbol prefix. A scalar variable can be either a number or a string.

2. Array

denoted by a @ symbol prefix. Arrays are indexed by numbers.

3. Associative Array

denoted by a % symbol prefix. Arrays are indexed by strings. You can look up items by name.

Note:	This is quite different than Pascal and even C. But Hopefully this makes things easier.

The different beginning characters help you understand how a variable is used when you look at someone else's Perl code. If you see a variable called @Value, you automatically know that it is an array variable.

They also provide a different namespace for each variable type. Namespaces separate one set of names from another. This lets you use $name, @name, and %name to refer to different values.

1. Scalar Variables

Scalar variables can be either a number or a string.

(i) You can use variable types almost interchangeably. Numbers first then strings later

(ii) Numbers are numbers there is no integer type. Perl regards all numbers a floating-point numbers for calculations etc.

(a) Defining Scalar Variables

You define scalar variables by assigning a value (number or string) to it.

(i) You do not declare variable types at a special place in the program as in PASCAL.

(ii) It is a good idea to declare all variables together near the top of the program.

The following are simple examples (var1.pl) of variable declarations :

```
$first_name = "Tanmay";
$last_name = "Arya";
$number = 3;
$another_number = 1.25;
$sci_number = 7.25e25;
$octal_number = 0377; # same as 255 decimal
$hex_number = 0xff; # same as 255 decimal
```

(b) String Scalar Variables

Strings are a sequence of characters. Perl has two types of string :

(i) Single-quoted strings

denoted by '....'. All characters are regarded as being literal characters.

That is to say special format characters like \n are regarded as being two characters \ and with no special meaning.

Two exceptions :

- To get a single-quote character do \'
- To get a backslash character do \\

(ii) Double-quoted strings

Special format characters now have a special meaning.

Some special format characters include :

```
\n   newline
\r   carriage return
\t   tab
\b   backspace
\\   backslash character
\"   double-quote character
\l   lower case next letter
\L   lower case all letters until \E
\u   upper case next letter
```

```
\U   upper case all letters until \E
\E   Terminate \L or \E
```

7.5.4. Arrays

We can have literal and variable arrays in Perl

We can also have two forms of array :

- **Arrays :** An array is a series of numbers and strings handled as a unit. You can also think of an array as a list.
- **Associative Arrays :** This is the most complicated data type. Think of it as a list in which every value has an associated lookup item.

Perl uses arrays-or lists-to store a series of items. You could use an array to hold all of the lines in a file, to help sort a list of addresses, or to store a variety of items.

An array is an ordered list of scalar data.

Each element of an array is an separate scalar variable with a independent scalar value -- unlike PASCAL or C.

Arrays can therefore have **mixed** elements, for example

(1,"fred", 3.5)

is perfectly valid.

We have different types of array in perl. which is given below :

1. Literal Arrays

Arrays can be defined literally in Perl code by simply enclosing the array elements in parentheses and separating each array element with a comma.

For example,

```
(1, 2, 3)
("fred", "albert")
() # empty array (zero elements)
(1..5) # shorthand for (1,2,3,4,5)
```

2. Indexed Arrays

You declare an ordinary indexed array by giving it a name and prefixing it with a @

Values are assigned to arrays in the usual fashion :

```
@array = (1,2,3);
@copy_array = @array;
@one_element = 1;
# not an error but create the array (1)
```

Arrays can also be referenced in the literal list.

For example :

```
@array1 = (4,5,6);
@array2 = (1,2,3, @array1, 7,8);
```

results in the elements of array1 being inserted in the appropriate parts of the list.

Therefore after the above operations :

```
@array2 = (1,2,3,4,5,6,7,8)
```

This means Lists cannot contain other lists elements only scalars allowed.

Elements of arrays are indexed by index :

```
$array1[1] = $array2[3];
```

Assign ``third" element of array2 to ``first" element of array1.

Each array element is a scalar so we reference each array element with $.

3. Associative Arrays

Associative arrays are a very useful and commonly used feature of Perl.

Associative arrays basically store tables of information where the lookup is the right hand key to an associated scalar value. Again scalar values can be mixed ``types".

Associative arrays are denoted by a verb| When you declare an associative array the key and associated values are listed in consecutive pairs.

So if we had the following secret code lookup :

name	code
Tanmay	1234
Anu	3456
Arya	6789

We would declare a Perl associative array to perform this lookup as follows :

```
%lookup = ("tanmay", 1234,
"Anu", 3456,
"Arya", 6789);
```

The reference a particular value you do :

```
$lookup{"Tanmay"}
```

You can create new elements by assignments to new keys. E.g.

```
$lookup{"Tanmay"} = 3845;
```

You do new assignments to old keys also :

```
# change dave's code
$lookup{"Tanmay"} = 7634;
```

Associative Array Operators

keys()

The keys(%arrayname) lists all the key names in a specified associative array. The answer is returned as an ordinary index array.

```
E.g.
@names = keys(%lookup);
```

values() This operator returns the values of the specified associative array.

```
E.g.
@codes = values(%lookup);
```

delete deletes an associated key and value by key reference, e.g.

```
# scrub Tanmay from code list
delete $lookup("Tanmay");
```

7.5.5. Operators

Perl has more operators than most languages. You've already seen some operators-like the equals or assignment operator(=). As you read about the other operators, you will undoubtedly realize that you are familiar with some of them.

Operators are instructions you give to the computer so that it can perform some task or operation. All operators cause actions to be performed on operands. An operand can be anything that you perform an operation on. In practical terms, any particular operand will be a literal, a variable, or an expression. You have already been introduced to literals and variables. A good working definition of expression is some combination of operators and operands that are evaluated as a unit.

Operands are also recursive in nature. In Perl, the expression 3 + 5-two operands and a plus operator-can be considered as one operand with a value of 8. For instance, (3 + 5) - 12 is an expression that consists of two operands, the second of which is subtracted from the first. The first operand is (3 + 5) and the second operand is 12.

1. The Binary Arithmetic Operators

There are six binary arithmetic operators: **addition**, **subtraction**, **multiplication**, **exponentiation** (**), **division**, and **modulus** (%).

For example, op1.pl :

```
$x = 3 + 1;
$y = 6 - $x;
$z = $x * $y;
$w = 2**3; # 2 to the power of 3 = 8
```

The modulus operator is useful when a program needs to run down a list and do something every few items. This example shows you how to do something every 10 items, mod.pl :

```
for ($index = 0; $index <= 100; $index++) {
    if ($index % 10 == 0) {
        print("$index ");
    }
}
```

When this program is run, the output should look like the following :

```
0 10 20 30 40 50 60 70 80 90 100
```

Notice that every tenth item is printed. By changing the value on the right side of the modulus operator, you can affect how many items are processed before the message is printed. Changing the value to 15 means that a message will be printed every 15 items.

2. The Unary Arithmetic Operators

In common with C, Perl has two short hand unary operators that can prove useful.

The unary arithmetic operators act on a single operand. They are used to change the sign of a value, to increment a value, or to decrement a value. Incrementing a value means to add one to its value. Decrementing a value means to subtract one from its value.

In many statements we frequently write something like :

```
$a = $a + 1;
```

we can write this more compactly as :

```
$a += 1;.
```

This works for any operator so this is equivalent :

```
$a = $a * $b;
$a *= $b;
```

You can also automatically increment and decrement variables in Perl with the ++ and -- operators.

For example all three expressions below achieve the same result :

```
$a = $a + 1;
$a += 1;
++$a;
```

The ++ and -- operators can be used in prefix and postfix mode in expressions. There is a difference in their use.

In Prefix the operator comes before the variable and this indicates that the value of the operation be used in the expression : I.e.

```
$a = 3;
$b = ++$a;
```

results in a being incremented to 4 before this new value is assigned to b. That is to say BOTH a and b have the value 4 after these statements have been executed.

In postfix the operator comes after the variable and this indicates that the value of the variable before the operation be used in the expression and then the variable is incremented or decremented: I.e.

```
$a = 3;
$b = $a++;
```

results in the value of a (3) being assigned to b and then a gets incremented to 4 That is to that after these statements have been executed b = 3 and a = 4.

The Perl programming language has many ways of achieving the same objective. You will become a more efficient programmer if you decide on one approach to incrementing/decrementing and use it consistently.

3. The Logical Operators

Logical operators are mainly used to control program flow. Usually, you will find them as part of an if, a while, or some other control statement.

The Logical operators are :

op1 && op2

⇒ Performs a logical AND of the two operands.

op1 || op2

⇒ Performs a logical OR of the two operands.

!op1

⇒ Performs a logical NOT of the operand.

The concept of logical operators is simple. They allow a program to make a decision based on multiple conditions. Each operand is considered a condition that can be evaluated to a true or false value. Then the value of the conditions is used to determine the overall value of the op1 operator op2 or !op1 grouping. The following examples demonstrate different ways that logical conditions can be used.

The && operator is used to determine whether both operands or conditions are true and.pl.

For example :

```
if ($firstVar == 10 && $secondVar == 9) {
    print("Error!");
};
```

If either of the two conditions is false or incorrect, then the print command is bypassed.

The || operator is used to determine whether either of the conditions is true.

For example :

```
if ($firstVar == 9 || $firstVar == 10) {
    print("Error!");
```

If either of the two conditions is true, then the print command is run.

Caution If the first operand of the || operator evaluates to true, the second operand will not be evaluated. This could be a source of bugs if you are not careful.

For instance, in the following code fragment :

```
if ($firstVar++ || $secondVar++) { print("\n"); }
```

variable $secondVar will not be incremented if $firstVar++ evaluates to true.

The ! operator is used to convert true values to false and false values to true. In other words, it inverts a value. Perl considers any non-zero value to be true-even string values. For example :

```
$firstVar = 10;
$secondVar = !$firstVar;
if ($secondVar == 0) {
    print("zero\n");
};
```

is equal to 0- and the program produces the following output :

```
zero
```

You could replace the 10 in the first line with "ten," 'ten,' or any non-zero, non-null value.

4. The Bitwise Operators

The bitwise operators are similar to the logical operators, except that they work on a smaller scale binary representations of data.

The following operators are available :

- **op1 & op2** : The AND operator compares two bits and generates a result of 1 if both bits are 1; otherwise, it returns 0.

- **op1 | op2** : The OR operator compares two bits and generates a result of 1 if the bits are complementary; otherwise, it returns 0.
- **op1^ op2** : The EXCLUSIVE-OR operator compares two bits and returns 1 if either of the bits are 1 and it gives 0 if both bits are 0 or 1.
- **~op1** : The COMPLEMENT operator is used to invert all of the bits of the operand.
- **op1 >> op2** : The SHIFT RIGHT operator moves the bits to the right, discards the far right bit, and assigns the leftmost bit a value of 0. Each move to the right effectively divides op1 in half.
- **op1 << op2** : The SHIFT LEFT operator moves the bits to the left, discards the far left bit, and assigns the rightmost bit a value of 0. Each move to the left effectively multiplies op1 by 2.

5. Comparison operators for numbers and strings

Perl has different operators for comparing numbers and strings. They are defined as follows :

Equality	Numeric	String
Equal	==	eq
Not Equal	!=	ne
Comparison	<=>	cmp
Relational	Numeric	String
Less than	<	lt
Greater than	>	gt
Less than or equal	<=	le
Greater than or equal	>=	ge

In controlling the logic of a conditional expression logical operators are frequently required. In Perl, The logical AND operator is && and logical OR is ||.

For example to check if a valid number exist in a variable $var you could do :

```
if ( ($var ne "0") && ($var == 0))
    { # $var is a number
    }
```

Since Perl evaluates any string to 0 if is not a number.

The numeric relational operators, listed above are used to test the relationship between two operands. You can see if one operand is equal to another, if one operand is greater than another, or if one operator is less than another.

6. The Range Operator (..)

The range operator is used as a shorthand way to set up arrays.

When used with arrays, the range operator simplifies the process of creating arrays with contiguous sequences of numbers and letters. We will start with an array of the numbers one through ten.

For example tp Create an array with ten elements that include 1, 2, 3, 4, 5, 6, 7, 8, 9, and 10 simply write :

```
@array = (1..10);
```

You can also create an array of contiguous letters. For example: an array with ten elements that include A, B, C, D, E, F, G, H, I , and J. is written

```
@array = ("A".."J");
```

You can have other things in the array definition besides the range operator.

You can create an array that includes AAA, 1, 2, 3, 4, 5, A, B, C, D, and ZZZ by the following :

```
@array = ("AAA", 1..5, "A".."D", "ZZZ");
```

You can use the range operator to create a list with zero-filled numbers.

To create an array with ten elements that include the strings 01, 02, 03, 04, 05, 06, 07, 08, 09, and 10 do :

```
@array = ("01".."10");
```

And you can use variables as operands for the range operator.

To assign a string literal to $firstVar. Create an array with ten elements that include the strings 01, 02, 03, 04, 05, 06, 07, 08, 09, and 10 :

```
$firstVar = "10";
@array = ("01"..$firstVar);
```

If you use strings of more than one character as operands, the range operator will increment the rightmost character by one and perform the appropriate carry operation when the number 9 or letter z is reached. You'll probably need to see some examples before this makes sense.

You've already seen "A".."Z," which is pretty simple to understand. Perl counts down the alphabet until Z is reached.

7. The String Operators (. and x)

Perl has two different string operators-the concatenation (.) operator and the repetition (x) operator. These operators make it easy to manipulate strings in certain ways. Let's start with the concatenation operator. Strings can be concatenated by the . operator.

For example :

```
$first_name = "Tanmay";
$last_name = "Arya";
$full_name = $first_name . " " . $last_name;
```

we need the " " to insert a space between the strings.

```
Strings can be repeated with tt x operator
```

For example :

```
$first_name = "Tanmay";
Tanmay_cubed = $first_name x 3;
```

which gives "TanmayTanmayTanmay".

String can be referenced inside strings

For example :

```
$first_name = "tanmay";
$str = "My name is: $first_name";
which gives "My name is: Tanmay".
```

8. Conversion between numbers and Strings

This is a useful facility in Perl. Scalar variables are converted automatically to string and number values according to context.

Thus you can do :

```
$x = "40";
$y = "11";
$z = $x + $y; # answer 51
$w = $x . $y; # answer "4011"
```

Note if a string contains any trailing non-number characters they will be ignored.

i.e. " 123.45abc" would get converted to 123.45 for numeric work.

If no number is present in a string it is converted to 0.

9. The chop() operator

chop() is a useful operator which takes a single argument (within parenthesis) and simply removes the last character from the string. The new string is returned to the input string.

Thus

```
chop('Hello') would give the result 'Hello'
```

Why is this useful?

Most strings input in Perl will end with a \n. If we want to string operations for output formatting and many other processed then the \n might be inappropriate. chop() can easily remove this.

Many other applications find chop() useful

7.5.6. Functions

Functions are blocks of codes that are given names so that you can use them as needed. Functions help you to organize your code into pieces that are easy to understand and work with. They let you build your program step by step, testing the code along the way.

If your program has a function that calculates the area of a circle, the following line of code might be used to call it :

```
$areaOfFirstCircle = areaOfCircle($firstRadius);
```

By looking at the function call, the reader knows what the program is doing. Detailed understanding of the actual function is not needed.

Here is an example call of the function, func1/pl :

```
$areaOfFirstCircle = areaOfCircle(5);
print("$areaOfFirstCircle\n");
sub areaOfCircle {
    $radius = $_[0];
    return(3.1415 * ($radius ** 2));
}
```

This program prints :

```
78.7375
```

The fact that something prints tells you that the program flow returned to the print line after calling the areaOfCircle() function.

1. Using the Parameter Array (@_)

All parameters to a function are stored in an array called @_. One side effect of this is that you can find out how many parameters were passed by evaluating @ in a scalar context.

For example, func2.pl:

```
firstSub(1, 2, 3, 4, 5, 6);
firstSub(1..3);
firstSub("A".."Z");
sub firstSub {
    $numParameters = @_ ;
    print("The number of parameters is $numParameters\n");
}
```

This program prints out :

```
The number of parameters is 6
The number of parameters is 3
The number of parameters is 26
```

Perl lets you pass any number of parameters to a function. The function decides which parameters to use and in what order. The @_ array is used like any other array.

2. Passing Parameters by Reference

Using scalar variables inside your functions is a good idea for another reason-besides simple readability concerns. When you change the value of the elements of the @ array, you also change the value of the parameters in the rest of the program. This is because Perl parameters are called by reference. When parameters are called by reference, changing their value in the function also changes their value in the main program.

For example, fanc3.pl:

```
@array = (0..5);
print("Before function call, array = @array\n");
firstSub(@array);
print("After function call, array = @array\n");
sub firstSub{
    $_[0] = "A";
    $_[1] = "B";
}
```

This program prints :

```
Before function call, array = 0 1 2 3 4 5
After function call, array = A B 2 3 4 5
```

You can see that the function was able to affect the @array variable in the main program. Generally, this is considered bad programming practice because it does not isolate what the function does from the rest of the program. If you change the function so that scalars are used inside the function, this problem goes away.

So we now write, fanc3.pl :

```
@array = (0..5);
print("Before function call, array = @array\n");
firstSub(@array);
print("After function call, array = @array\n");
sub firstSub{
     ($firstVar, $secondVar) = @_ ;
     $firstVar = "A";
     $secondVar = "B";
}
```

This program prints :

```
Before function call, array = 0 1 2 3 4 5
After function call, array = 0 1 2 3 4 5
```

This example shows that the original @array variable is left untouched. However, another problem has quietly arisen. Let's change the program a little so the values of $firstVar are printed before and after the function call.

3. Scope of Variables

Scope refers to the visibility of variables. In other words, which parts of your program can see or use it. Normally, every variable has a global scope. Once defined, every part of your program can access a variable.

It is very useful to be able to limit a variable's scope to a single function. In other words, the variable wil have a limited scope. This way, changes inside the function can't affect the main program in unexpected ways, func6.pl

```
firstSub("AAAAA", "BBBBB");
sub firstSub{
     local ($firstVar) = $_[0];
     my($secondVar) = $_[1];
     print("firstSub: firstVar = $firstVar\n");
     print("firstSub: secondVar = $secondVar\n\n");
     secondSub();
     print("firstSub: firstVar = $firstVar\n");
     print("firstSub: secondVar = $secondVar\n\n");
}
sub secondSub{
     print("secondSub: firstVar = $firstVar\n");
     print("secondSub: secondVar = $secondVar\n\n");
     $firstVar = "ccccC";
     $secondVar = "DDDDD";
```

```
    print("secondSub: firstVar = $firstVar\n");
    print("secondSub: secondVar = $secondVar\n\n");
}
```

This program prints :

```
firstSub: firstVar = AAAAA
firstSub: secondVar = BBBBB
secondSub: firstVar = AAAAA
Use of uninitialized value at test.pl line 19.
secondSub: secondVar =
secondSub: firstVar = CCCCC
secondSub: secondVar = DDDDD
firstSub: firstVar = CCCCC
firstSub: secondVar = BBBBB
```

The output from this example shows that secondSub() could not access the $secondVar variable that was created with my() inside firstSub(). Perl even prints out an error message that warns about the uninitialized value. The $firstVar variable, however, can be accessed and valued by secondSub().

4. Using a List as a Function Parameter

Now that you understand about the scope of variables, let's take another look at parameters. Because all parameters are passed to a function in one array, what if you need to pass both a scalar and an array to the same function? This next example shows you what happens, func7.pl

```
firstSub((0..10), "AAAA");
sub firstSub{
    local(@array, $firstVar) = @_ ;
    print("firstSub: array = @array\n");
    print("firstSub: firstVar = $firstVar\n");
}
```

This program prints :

```
firstSub: array = 0 1 2 3 4 5 6 7 8 9 10 AAAA
Use of uninitialized value at test.pl line 8.
firstSub: firstVar =
```

When the local variables are initialized, the @array variables grab all of the elements in the @ array, leaving none for the scalar variable. This results in the uninitialized value message displayed in the output. You can fix this by merely reversing the order of parameters. If the scalar value comes first, then the function processes the parameters without a problem, func8.pl

```
firstSub("AAAA", (0..10));
sub firstSub{
    local($firstVar, @array) = @_ ;
    print("firstSub: array = @array\n");
```

```
    print("firstSub: firstVar = $firstVar\n");
}
```

This program prints :

```
firstSub: array = 0 1 2 3 4 5 6 7 8 9 10
firstSub: firstVar = AAAA
```

5. Nesting Function Calls

Function calls can be nested many levels deep. Nested function calls simply means that one function can call another which in turn can call another. Exactly how many levels you can nest depends on which version of Perl you are running and how your machine is configured. Normally, you don't have to worry about it. If you want to see how many levels your system can recurse, try the following small program, func9.pl :

```
firstSub();
sub firstSub{
    print("$count\n");
    $count++;
    firstSub();
}
```

A system typically counts up to something like 127 before displaying the following message:

Error : Runtime exception

While it is important to realize that there is a limit to the number of times your program can nest functions, you should never run into this limitation unless you are working with recursive mathematical functions.

6. Using a Private Function

Occasionally, you might want to create a private function. A private function is one that is only available inside the scope where it was defined, private.pl :

```
$temp = performCalc(10, 10);
print("temp = $temp\n");
sub performCalc {
    my ($firstVar, $secondVar) = @_;
    my $square = sub {
    return($_[0] ** 2);
    };
    return(&$square($firstVar) + &$square($secondVar));
};
```

This program prints :

```
temp = 200
```

This example is rather trivial, but it serves to show that in Perl it pays to create little helper routines. A fine line needs to be drawn between what should be included as a private function and what shouldn't. I would draw the line at 5 or 6 lines of code. Anything longer probably should be made into its own function. I would also say that a private function should have only one purpose for existence. Performing a calculation and then opening a file is too much functionality for a single private function to have.

The rest of the chapter is devoted to showing you some of the built-in functions of Perl. These little nuggets of functionality will become part of your arsenal of programming weapons.

7. String Functions

Perl provides a very useful set of string handling functions :

The first set of functions that we'll look at are those that deal with strings. These functions let you determine a string's length, search for a sub-string, and change the case of the characters in the string, among other things.

Here are Perl's string functions :

- **chomp(STRING) OR chomp(ARRAY) :** Uses the value of the $/ special variable to remove endings from STRING or each element of ARRAY. The line ending is only removed if it matches the current value of $/.
- **chop(STRING) OR chop(ARRAY) :** Removes the last character from a string or the last character from every element in an array. The last character chopped is returned.
- **chr(NUMBER) :** Returns the character represented by NUMBER in the ASCII table. For instance, chr(65) returns the letter A.
- **crypt(STRING1, STRING2) :** Encrypts STRING1. Unfortunately, Perl does not provide a decrypt function.
- **index(STRING, SUBSTRING, POSITION) :** Returns the position of the first occurrence of SUBSTRING in STRING at or after POSITION. If you don't specify POSITION, the search starts at the beginning of STRING.
- **join(STRING, ARRAY) :** Returns a string that consists of all of the elements of ARRAY joined together by STRING. For instance, join(">>", ("AA", "BB", "cc")) returns "AA>>BB>>cc".
- **lc(STRING) :** Returns a string with every letter of STRING in lowercase. For instance, lc("ABCD") returns "abcd".
- **lcfirst(STRING) :** Returns a string with the first letter of STRING in lowercase. For instance, lcfirst("ABCD") returns "aBCD".
- **length(STRING) :** Returns the length of STRING.
- **rindex(STRING, SUBSTRING, POSITION) :** Returns the position of the last occurrence of SUBSTRING in STRING at or after POSITION. If you don't specify POSITION, the search starts at the end of STRING.
- **split(PATTERN, STRING, LIMIT) :** Breaks up a string based on some delimiter. In an array context, it returns a list of the things that were found. In a scalar context, it returns the number of things found.
- **substr(STRING, OFFSET, LENGTH) :** Returns a portion of STRING as determined by the OFFSET and LENGTH parameters. If LENGTH is not specified, then everything from OFFSET to the end of STRING is returned. A negative OFFSET can be used to start from the right side of STRING.
- **uc(STRING) :** Returns a string with every letter of STRING in uppercase. For instance, uc("abcd") returns "ABCD".

- **Ucfirst(STRING) :** Returns a string with the first letter of STRING in uppercase. For instance, ucfirst("abcd") returns "Abcd".

8. Array Functions

Arrays are also a big part of the Perl language and Perl has a lot of functions to help you work with them. Some of the actions arrays perform include deleting elements, checking for the existence of an element, reversing all of the the elements in an array, and sorting the elements.

Here are the functions you can use with arrays :

- **defined(VARIABLE) :** Returns true if VARIABLE has a real value and if the variable has not yet been assigned a value. This is not limited to arrays; any data type can be checked. Also see the exists function for information about associative array keys.
- **delete(KEY) :** Removes the key-value pair from the given associative array. If you delete a value from the %ENV array, the environment of the current process is changed, not that of the parent.
- **each(ASSOC_ARRAY) :** Returns a two-element list that contains a key and value pair from the given associative array. The function is mainly used so you can iterate over the associate array elements. A null list is returned when the last element has been read.
- **exists(KEY) :** Returns true if the KEY is part of the specified associative array. For instance, exists($array{"Orange"}) returns true if the %array associative array has a key with the value of "Orange."
- **join(STRING, ARRAY) :** Returns a string that consists of all of the elements of ARRAY joined together by STRING. For instance, join(">>", ("AA", "BB", "cc")) returns "AA>>BB>>cc".
- **keys(ASSOC_ARRAY) :** Returns a list that holds all of the keys in a given associative array. The list is not in any particular order.
- **map(EXPRESSION, ARRAY) :** Evaluates EXPRESSION for every element of ARRAY. The special variable $ is assigned each element of ARRAY immediately before EXPRESSION is evaluated.
- **pack(STRING, ARRAY) :** Creates a binary structure, using STRING as a guide, of the elements of ARRAY. You can look in the next Chapter on References for more information.
- **pop(ARRAY) :** Returns the last value of an array. It also reduces the size of the array by one.
- **push(ARRAY1, ARRAY2) :** Appends the contents of ARRAY2 to ARRAY1. This increases the size of ARRAY1 as needed.
- **reverse(ARRAY) :** Reverses the elements of a given array when used in an array context. When used in a scalar context, the array is converted to a string, and the string is reversed.
- **scalar(ARRAY) :** Evaluates the array in a scalar context and returns the number of elements in the array.
- **shift(ARRAY) :** Returns the first value of an array. It also reduces the size of the array by one.
- **sort(ARRAY) :** Returns a list containing the elements of ARRAY in sorted order. See next Chapter 8on References for more information.
- **splice(ARRAY1, OFFSET, :** Replaces elements of ARRAY1 with elements LENGTH, ARRAY2)in ARRAY2. It returns a list holding any elements that were removed. Remember

that the $[variable may change the base array subscript when determining the OFFSET value.

- **split(PATTERN, STRING, LIMIT) :** Breaks up a string based on some delimiter. In an array context, it returns a list of the things that were found. In a scalar context, it returns the number of things found.
- **undef(VARIABLE) :** Always returns the undefined value. In addition, it undefines VARIABLE, which must be a scalar, an entire array, or a subroutine name.
- **unpack(STRING, ARRAY) :** Does the opposite of pack().
- **unshift(ARRAY1, ARRAY2) :** Adds the elements of ARRAY2 to the front of ARRAY1. Note that the added elements retain their original order. The size of the new ARRAY1 is returned.
- **values(ASSOC_ARRAY) :** Returns a list that holds all of the values in a given associative array. The list is not in any particular order.

7.5.7. If/Unless statement

The if statement has a variety of forms. The simplest is :

```
if (expression)
    { true_statement_1;
    .....
    true_statement_n;
    }
```

which means that if expression is evaluated as being true then execute the block of statements.

In Perl, false is regarded as any expression which evaluates to 0. true any expression which is not false (non-zero).

An else may be added to provide a block of statements to be executed upon a false evaluation of the expression :

```
if (expression)
    { true_statement_1;
    .....
    true_statement_n;
    }
else
    { false_statement_1;
    .....
    false_statement_n;
    }
```

Curly braces are required for each block even if only one statement is present.

For example :

```
$age = ; # whatever ??
if ($age < 18)
    { print "You cannot Vote or have a beer, yet.\n";
```

```
    }
else
    { print "Go and Vote and then have a beer.\n";
    }
```

There is an unless statement in Perl which can be regarded as the negative of if: If the control expression is not true, do ....

```
$age = ; # whatever ??
unless ($age < 18)
    { print "Go and Vote and then have a beer.\n";
    }
```

unless can have an else, too.

If you have more than one branch then the elsif can be added to the if. You cannot have an else if in Perl -- you must use elsif.

```
$age = ; # whatever ??
if ($age < 16)
    {
    print "Hi, Junior\n";
    }
elsif ($age < 17)
    {
    print You can ????\n";
    }
elsif ($age < 18)
    {
    print You can learn to drive\n";
    }
else
    { print "Go and Vote and then have a beer.\n";
    }
```

7.5.8. The for statement

Just like in PASCAL we have loops.

The simplest is the for loop. It actually behaves like C's statement and is a little more complex than PASCAL -- it can do some more things.

The format of the for statement is :

```
for ( initialise_expr; test_expr; increment_expr )
    {
    statement(s);
```

```
    }
```

For example

```
for ( $i = 1; $i <= 10; ++$i )
    { # count to 10
    print "$i\n";
    }
```

7.5.9. The while/until statement

The while statement is as follows :

```
while (expression)
    { # while expression is true execute this block
    statement(s);
    }
```

For example (count_while.pl) :

```
i = 1;
while ( $i <= 10 )
    { # count to 10
    print "$i\n";
    ++$i;
    }
```

The until statement says do while expression is false as declared is as follows :

until (expression)

```
{ # until expression is false execute this block
    statement(s);
}
```

For example (count_until.pl) :

```
i = 1;
until ( $i > 10 )
    { # count to 10
    print "$i\n";
    ++$i;
    }
```

7.5.10. The foreach statement

The foreach statement iterates through items in a list: The statement has the following format :

```
foreach $i (@some_list)
    { # $i takes on each list item value in turn
    statement(s);
```

```
    }
```

For example (foreach.pl) :

```
@a = (1, 2, 3, 4, 5);
foreach $i (reverse @a)
    { # reverse is a functiion that flips the list order
    print $i;
    }
```

☞ SUMMARY

This chapter has described the basic concepts of the grep, sed filter as well as awk, perl advance filter and given a short introduction of concept of awk functions like as user defined function and built-in function. At this point you should be able to write short but very useful awk programs to perform many tasks.

This chapter is only an introduction to awk. It should be enough to give you a sense of the language and its potential and make it possible for you to learn more by using it.

☞ EXERCISE

1. What do you mean by awk ? Explain the awk in details.
2. What is pattern? Explain different types of patterns.
3. What does the -f option tell the awk utility?
4. How we can run the awk program? Explain with example.
5. Explain different built-in functions.
6. What is the default if the pattern part of a rule is missing?
7. What is the difference between AWK and SED?
8. What is the syntax of grep?
9. What is SED?
10. What is grep? Explain in details.
11. What do you mean by perl? Define in details.
12. What is the purpose of the grep?

❑❑❑

CHAPTER 8

UNIX TOOLS

8.1 INTRODUCTION

The UNIX operating system is very popular operating system for multi-user in different fields. As we know that in multi-user operating system several user use the same program (system) at the same time. As a result of this exposure UNIX system becomes commercial software with helpful application tool and system development tools. UNIX system contains different tools that provide a complete solution for text formatting, system development and many more commercial applications

UNIX system operating system support different types of tools, which is given below:

8.2 LANGUAGE DEVELOPMENT TOOLS

8.2.1 Yacc-Yet another Compiler Compiler a parser generator

Syntax

```
$yacc [option ] filename
```

The yacc command is used to generate C code that implements a parser. The program that yacc generates is based on an input file consisting of a grammar and possibly some C code fragments and internal declarations. yacc converts a context-free grammar into a set of tables for a simple automaton which executes a parsing algorithm. Code that is executed whenever a grammar production is recognized may be supplied. The grammar may be ambiguous; precedence rules are used to resolve ambiguities.

The output file containing the parser code is called y.tab.c. This program must be compiled by the C compiler; the parsing routine is called yyparse.

Firstly a YACC specification for an expression is created. The YACC compiler transfers the file parse.y into a C program called y.tab.c. The C program y.tab.c is run through C compiler and produce object program that perform the translation specified by the original YACC program

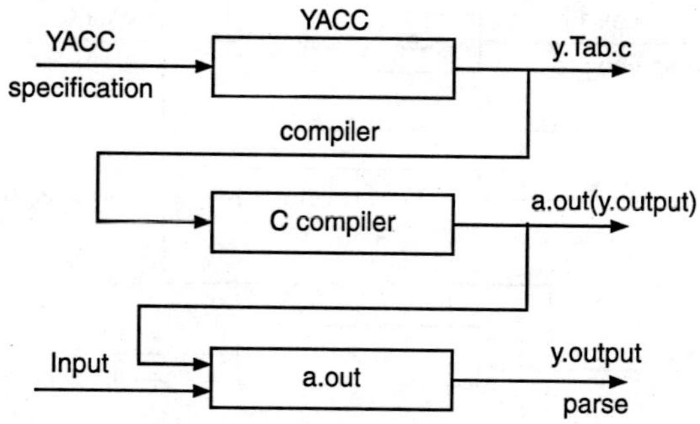

FIGURE 8.1 YACC diagram

Options

-d A file called y.tab.c will be generated. This file may be included in other programs which need to access the token codes.

-l This option should only be used after the grammar and the associated actions are fully debugged.

-S Silently ignored for backwards compatibility.

-t Runtime debugging code is always generated in y.tab.c.

-v A file called y.output will be generated.

-V Print out version information on standard error.

The number of reduce-reduce and shift-reduce conflicts is reported on the standard error output; a more detailed report is found in the y.output file. If some rules are not reachable from the start symbol, this is reported.

8.2.2 lex- Lexical Analyzer Generator

Syntax :

```
$lex [option][file name]
```

The **lex** command generates C code which implements a lexical analyzer—a routine which reads input text and separates it into tokens. The **lex** command assist in the development of lexical analyzer of the compiler. Lex command has been widely used to specify lexical analyzer for variety of languages.

Lex is a software tool that takes as input a specification of a set of regular expressions together with actions to be taken on recognizing each of these expressions. **lex** generates a file of C code called *lex.yy.c*. This file must be compiled by the C compiler and linked with a **main** routine. The program should be linked with the **lex** library. This library supplies a **main** routine. The lexical analyzer routine produced is called **yylex**. This routine reads its input and, when a token is recognized, executes the code associated with the token class. The default action is to write the token to the standard input.

The regular expressions understood by **lex** contain many of the usual operators and special characters.

First lexical analyzer creates a program lex.1 in the lex language. Then lex.1 runs through lex compiler to produce a c program lex.y.y.c. lex.y.y.c is run through the c compiler to produce an object code which is a lexical analyzer that transforms an input to a sequence of tokens.

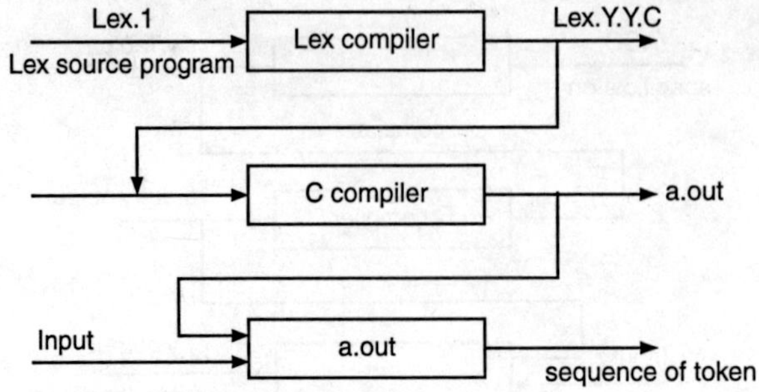

FIGURE 8.2 Lex diagram

Options

The options must appear before any files.

-c Indicates C actions and is the default.

-t This causes the generated code to be written to the standard out rather than to *lex.yy.c*

-v Provides a one-line summary of statistics.

-n Suppresses the summary of statistics even if **-v** is turned on.

-V Print out version information on standard error.

1. Compiler Design

Before design of the compiler one question arrive in my mind that, what is compiler? A compiler is a software or program that read a program written in a source language and translates it into equivalent program in another language means target language. An assembler which translates assembly language into machine language. The important aspect of compilation is to produce error messages in the source program. The error messages are mainly due to the grammatical/syntactical mistakes done by the programmer.

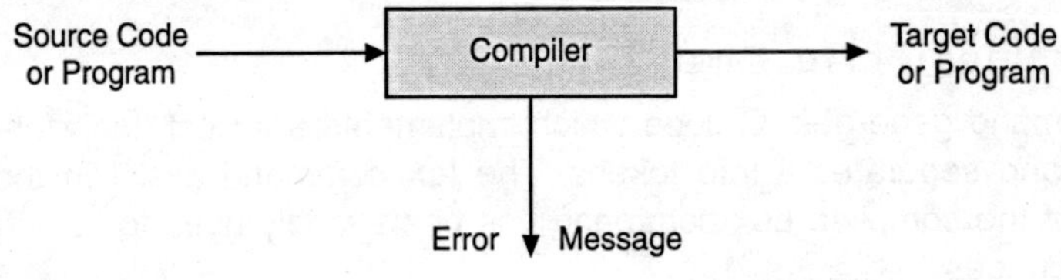

2. Compiler Design Phase

As we know that compiler being a complex program the software is developed through several phases. But we don't know about total number of phase of compiler. The working of compiler is divided broadly into two phases. In the compiler each phase transforms the source program from one representation to another. The phases of the compilers are:

1. Analysis phase
   - Lexical analysis
   - Syntax analysis
   - Semantic analysis
2. Synthesis phase

   Code optimization and Code generation

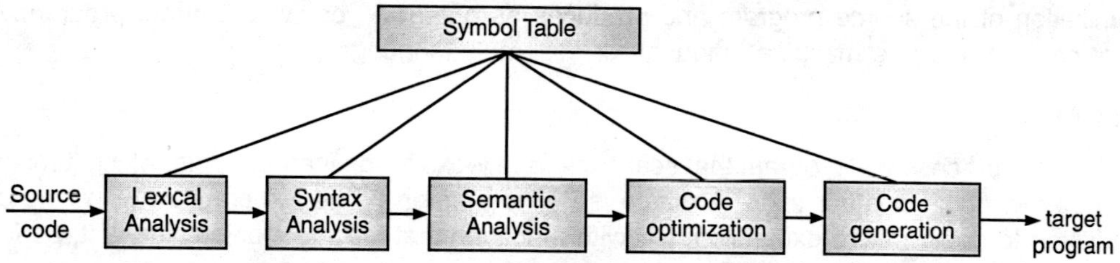

FIGURE 8.3

3. Lexical Analysis

Lexical analysis is the first phase of compiler. Lexical analysis scans a source code or program from left to right and character by character from left to right and break them into tokens. Each token or basic syntactic element represents a logical sequence of characters such as keyword, identifier, variables, and multi character operator etc. The output of lexical analysis phase goes to the next phase, i.e. parsing or syntax analysis.

4. Syntax Analysis

Every language whether it is any natural language or computer language or programming language follows certain grammatical rules that define syntactical structures of a language. For example a programming language like C language is made out of main functions consisting of blocks, blocks out of statements, statement out of expressions, and expression out of tokens and so on. The syntax of a programming language can be described by BNF (Backens Naur Form). These types of notations are also called context free grammars. As we know that syntax analysis is the second phase of the compiler phase, this is also called as parsing. The output of parsing is a representation of the syntactic structure of a statement in the form of parse tree.

The out put of syntax analysis is in the form of a tree is referred as parse tree.

For example, the statement A = B + C.

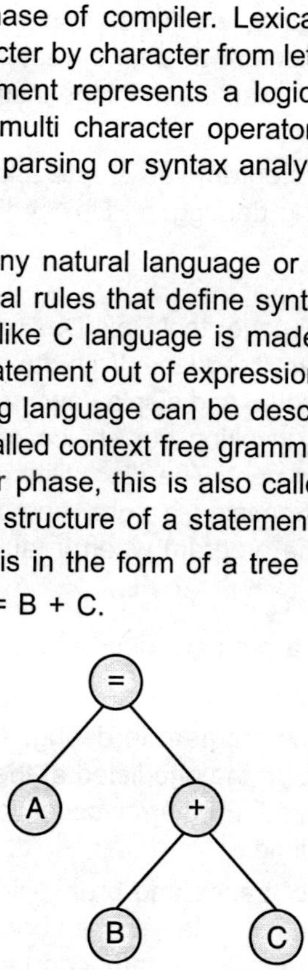

FIGURE 8.4 Parse tree

5. Semantic Analysis

The semantic analysis phase checks the source code or program for semantic errors. An important component of semantic analysis is type checking.

6. Synthesis Phase

The synthesis phase is considered as a phase of code generation and code optimization. The final phase of the compiler is the code generator. The code generator takes an input as code optimization of the source program and produces as output an equivalent target program. The target program may be machine language or assembly language.

8.2.3 M4

A macro processor is program that scans text and looks for defined symbols, which it replaces with other symbols or other texts. A macro is defined either by the M4 program or by the user. In addition to doing macro expansion, M4 with functions that include manipulates text, performs integer arithmetic, and any other files.

8.3 DEBUGGER TOOLS

When running and testing our programs, we like to use a debugger, which lets us single-step through a program, execute one statement at a time, restart the program, examine and modify variables, examine the runtime stack, examine machine instructions, assemble new instructions, and run a program up to a breakpoint.

Depending on the type of target program you create and your current operating system, We can select some or all of the following debuggers: Dbx, Adb, Sdb, Strip, and Ctrace.

8.3.1 Dbx

UNIX system have a powerful debugger tool Dbx. Dbx provides a powerful tool for monitoring the execution of programs written in any language. With dbx, we can suspend program execution, continue execution, skip sections of code, and display and set the values of variables. In addition, we can manipulate external files by invoking our choice of editors and issue shell commands without exiting the debugging environment. The dbx utility can be used to examine corefiles as well as executable programs. By examining a core memory image file, we can use the dbx command to examine the state of the program when it faulted.

Using Dbx

To use dbx to determine where a program died after a core dump, run

```
$dbx program
```

where program is the name of the program to debug. At the (dbx) prompt, say where. This will list the active functions on the stack; the one listed at the top (along with the exact line in the source code) is where the program died. Further investigation of the value of particular variables and such can also be done at that time.

Dbx also provides for a variety of traces and breakpoints, line-by-line and call-by-call step-throughs, investigation and manipulation of variables, memory, registers, signals and signal handlers, the stack, and more. Further documentation on dbx can be found in the on-line manual pages, hard copy documentation, and books written on the subject.

Dbx includes the following features:

1. Tracing.
2. Searching for functions.
3. Single tracing.
4. Editing from within the debugger.
5. Breaks points.

8.3.2 Adb (Absolute Debugger)

Adb is highly cryptic and difficult to use, still adb remains a popular debugger for the core programmers and is available in every version of UNIX. It is primarily an assembly level debugger and is not widely used.

The new version of the ADB debugger tool also provides better overall performance and reliability, including improvements to command line syntax and options, format strings, writing and searching, and expressions. The ADB Debugger also supports the following new features:

- Shared library support
- Threads support
- Multiprocessor dump reading support
- 64-bit DLKM dump reading support
- Single binary for 32-bit and 64-bit

To specify the new **adb** command, use the -n command line option.

For example,

```
$ adb -n /stand/vmunix /dev/kmem
    adb> runningprocs/d
        runningprocs:
            4
    adb> q
$
```

8.3.3 Sdb (Symbolic Debugger)

Sdb is also called symbolic debugger. It is available with virtual systems and debugs at a source level. It is used for checking executable C, assembly and Fortran programs.

The syntax of sdb command is:

```
$sdb [option] [object file ] [ corefile or dicrectory]
```

The features of Sdb command are:

1. Stack trace for C programs.
2. Knows about structures and arrays.
3. Breaking debugging.
4. Searching for char, float, and int point's patterns.
5. Examination of arbitrary files with no limit on size.
6. Debugging for c.

8.3.4 strip

Syntax:

```
$strip [option] filename
```

Strip is a debugger tool. The strip command strips the symbol table, debugging information, and line number information from end of line object files. It breaks the source code in defined sets of strips on the basis of defined numbers or logical linking code. Once this stripping process has been done, no symbolic debugging access will be available for that file; therefore, this command is normally run only on production modules that have been debugged and tested.

If strip is executed on a common archive file in addition to processing the members, strip will remove the archive symbol table. Using any of the following options can control the amount of information stripped from the end of file object file:

-l Strip line number information only.

-V Print standard error, the version number of strip.

-x Do not strip the symbol table.

strip is used to reduce the file storage overhead taken by the object file.

8.3.5 Ctrace - C program debugger

Syntax

```
$ctrace [options] [file]
```

The **ctrace** command allows us to follow the execution of a C program, statement-by-statement. It is not as powerful a symbolic debugger as Dbx or Sdb. The effect is similar to executing a shell procedure with the **-x** option. **ctrace** reads the C program in file . We must put the output of **ctrace** into a temporary file because **cc** does not allow the use of a pipe. We then compile and execute this file.

As each statement in the program executes, it will be listed at the terminal, followed by the name and value of any variables referenced or modified in the statement, followed by any output from the statement.

Options

The options commonly used are:

-f functions

Trace only these functions.

-v functions

Trace all but these functions.

We may want to add to the default formats for printing variables. Long and pointer variables are always printed as signed integers. Pointers to character arrays are also printed as strings if appropriate. **char, short,** and **int** variables are also printed as signed integers and, if appropriate, as characters. **double** variables are printed as floating-point numbers in scientific notation.

We can request that variables be printed in additional formats, if appropriate, with these options:

-o Octal

-x Hexadecimal

-u Unsigned

-e Floating point

Warning

We will get a **ctrace** syntax error if we omit the semicolon at the end of the last element declaration in a structure or union, just before the right brace (}). This is optional in some C compilers.

Defining a function with the same name as a system function may cause a syntax error if the number of arguments is changed. Just use a different name.

8.4 LIBRARY FUNCTIONS

The UNIX system provides a large number of C functions as libraries. Some of these implement frequently used operations, while others are very specialised in their application.

Wise programmers will check whether a library function is available to perform a task before writing their own version. This will reduce program development time. The library functions have been tested, so they are more likely to be correct than any function which the programmer might write. This will save time when debugging the program

8.4.1 Finding Information about Library Functions

The UNIX manual has an entry for all available functions. If you already know the name of the function you want, you can read the page by typing (to find about strcat).

```
$man 3 strcat
```

If you don't know the name of the function, a full list is included in the introductory page. To read this, type

```
$man 3 intro
```

There are approximately 700 functions described here. This number tends to increase with each upgrade of the system.

On any manual page, the SYNOPSIS section will include information on the use of the function. For example

```
#include <time.h>
char *ctime(time_t *clock)
```

This means that you must have

```
#include <time.h>
```

in your file before you call ctime. And that function ctime takes a pointer to type time_t as an argument, and returns a string (char *). time_t will probably be defined in the same manual page.

The DESCRIPTION section will give a short description of what the function does. For example

```
ctime() converts a long integer, pointed to by clock, to a
26-character string of the form produced by asctime().
```

8.4.2 Use of Library Functions

To use a function, ensure that you have made the required #includes in your C file. Then the function can be called as though you had defined it yourself.

It is important to ensure that your arguments have the expected types, otherwise the function will probably produce strange results. Lint is quite good at checking such things.

Some libraries require extra options before the compiler can support their use.

For example, to compile a program including functions from the math.h library the command might be

```
cc mathprog.c -o mathprog -m
```

The final -m is an instruction to link the maths library with the program. The manual page for each function will usually inform you if any special compiler flags are required.

8.4.3 Some Useful Library Functions

The following functions may be useful to you. Each manual page typically describes several functions, so if you see something similar to what you want, try looking in that manual page.

Function	Description
abs	integer absolute value
ctime	convert date and time
fopen	open a stream
printf	formatted output
fputc	put character or word on a stream
getwd	get current working directory path name
strcat	string concatenation
ctype	character classification and conversion macros and functions
mktemp	make a unique file name
puts	put a string on a stream
sleep	suspend execution, for interval
stdio	Standard buffered input/output package

8.5 SYSTEM CALLS

This section gives information about the system calls that interface with the UNIX operating system, such as open for opening a file, and exec for executing a program file. These are often accessed by C programmers.

A system call is just what its name implies — a request for the operating system to do something on behalf of the user's program. The system calls are functions used in the kernel itself. To the programmer, the system call appears as a normal C function call. However since a system call executes code in the kernel, there must be a mechanism to change the mode of a process from user mode to kernel mode. The C compiler uses a predefined library of functions that have the names of the system calls.

The library functions typically invoke an instruction that changes the process execution mode to kernel mode and causes the kernel to start executing code for system calls. The instruction that causes the mode change is often referred to as an "operating system trap" which is a software generated interrupt.

UNIX system calls are used to manage the file system, control processes, and to provide interprocess communication. The UNIX system interface consists of about 80 system calls.

8.5.1 Some Useful System Calls

The following table lists about 40 of the more important system call:

GENERAL CLASS	SPECIFIC CLASS	SYSTEM CALL
File Structure Related Calls	Creating a Channel	creat()
		open()
		close()
	Input/Output	read()
		write()
	Random Access	lseek()
	Channel Duplication	dup()
	Aliasing and Removing Files	link()
		unlink()
	File Status	stat()
		fstat()
	Access Control	access()
		chmod()
		chown()
		umask()
	Device Control	dctl()
Process Related Calls	Process Creation and Termination	exec()
		fork()
		wait()
		exit()
	Process Owner and Group	getuid()
		getgid()
	Process Identity	getpid()
		getppid()
	Process Control	signal()
		kill()
		alarm()
	Change Working Directory	chdir()
Interprocess Communication	Pipelines	pipe()
	Messages	msgget()
		msgsnd()
		msgrcv()
		msgctl()
	Semaphores	semget()
		semop()
	Shared Memory	shmget()

shmat()
shmdt()

When a system call discovers and error, it returns -1 and stores the reason the called failed in an external variable named "errno". The "/usr/include/errno.h" file maps these error numbers to manifest constants, and it these constants that you should use in your programs. When a system call returns successfully, it returns something other than -1, but it does not clear "errno". "errno" only has meaning directly after a system call that returns an error. When you use system calls in your programs, you should check the value returned by those system calls. Furthermore, when a system call discovers an error, you should use the "perror()" subroutine to print a diagnostic message on the standard error file that describes why the system call failed. The syntax for "perror()" is: void perror(string) char string; "perror()" displays the argument string, a colon, and then the error message, as directed by "errno", followed by a newline. The output of "perror()" is displayed on "standard error". Typically, the argument give to "perror()" is the name of the program that incurred the error, argv[0]. However, when using subroutines and system calls on files, the related file name might be passed to "perror()". There are occasions where you the programmer might wish to maintain more control over the printing of error messages than "perror()" provides—such as with a formatted screen where the newline printed by "perror()" would destroy the formatting.

Some UNIX system calls in details are as follows :

1. creat():Creating a Channel

All input and output operations start by opening a file using either the creat() or open() system calls. The prototype for the creat() system call is :

```
int creat(file_name, mode)
char *file_name;
int mode;
```

Where file_name is pointer to a null terminated character string that names the file and mode defines the file's access permissions. The mode is usually specified as an octal number such as 0666 that would mean read/write permission for owner, group, and others or the mode may also be entered using manifest constants defined in the "/usr/include/sys/stat.h" file. If the file named by file_name does not exist, the UNIX system creates it with the specified mode permissions.

However, if the file does exist, its contents are discarded and the mode value is ignored. The permissions of the existing file are retained.

For example :

```
/* creat.c */
#include<stdio.h>
#include<sys/types.h>      /*defines types used by sys/stat.h*/
   #include<sys/stat.h>    /*defines S_IREAD &amp; S_IWRITE*/
int main()
   {
      int fd;
      fd = creat("datafile.text", S_IREAD | S_IWRITE);
```

```
        if (fd == -1)
            printf("Error in opening datafile.text\n");
        else
            {
            printf("datafile.text opened for read/write access\n");
            printf("datafile.text is currently empty\n");
            }
        close(fd);
        exit (0);
    }
```

Multiple mode values may be combined by using the | operator the values together as demonstrated in the above sample program.

2. open() : open a file

Next file structure related system call is the open().Lets you open a file for reading, writing, or reading and writing.

The prototype for the open() system call is :

```
#include<fcntl.h>
int open(file_name, option_flags [, mode])
    char *file_name;
    int option_flags, mode;
```

Where file_name is a pointer to the character string that names the file, option_flags represent the type of channel, and mode defines the file's access permissions if the file is being created.

Multiple values are combined using the | operator (i.e. bitwise OR).

The mode argument may be specified in the same manner as in the creat() system call.

For example :

```
/* open.c */
#include<fcntl.h>       /* defines options flags */
#include<sys/types.h> /* defines types used by sys/stat.h */
#include<sys/stat.h>  /* defines S_IREAD &amp; S_IWRITE */
static char message[] = "Hello, world";

    int main()
    {
        int fd;
        char buffer[80];

            /* open datafile.dat for read/write access (O_RDWR)
                create datafile.dat if it does not exist (O_CREAT)
```

```
            return error if datafile already exists (O_EXCL)
            permit read/write access to file (S_IWRITE|S_IREAD)
          */
  fd=open("datafile.text",O_RDWR|O_CREAT|O_EXCL,S_IREAD|S_IWRITE);
    if (fd != -1)
       {
       printf("datafile.text opened for read/write access\n");
       write(fd, message, sizeof(message));
       lseek(fd, 0L, 0);  /*go back to the beginning of the file*/
       if (read(fd, buffer, sizeof(message)) == sizeof(message))
          printf("\"%s\" was written to datafile.text\n", buffer);
       else
          printf("error reading datafile.text\n");
       close (fd);
       }
    else
       printf(" datafile.text already exists \n");
    exit (0);
  }
```

3. close() : close a file

To close a channel which is use the close() system call. The prototype for the close() system call is :

```
int close(file_descriptor)
int file_descriptor;
```

Where file_descriptor identifies a currently open channel. close() fails if file_descriptor does not identify a currently open channel.

4. read() and write() : Input/Output

The read() system call does all input and the write() system call does all output. When used together, they provide all the tools necessary to do input and output sequentially. When used with the lseek() system call, they provide all the tools necessary to do input and output randomly.

Both read() and write() take three arguments. Their prototypes are :

```
int read(file_descriptor, buffer_pointer, transfer_size)
int file_descriptor;
char *buffer_pointer;
unsigned transfer_size;

int write(file_descriptor, buffer_pointer, transfer_size)
int file_descriptor;
```

```
char *buffer_pointer;
unsigned transfer_size;
```

Where file_descriptor identifies the I/O channel, buffer_pointer points to the area in memory where the data is stored for a read() or where the data is taken for a write(), and transfer_size defines the maximum number of characters transferred between the file and the buffer.

The read() and write() return the number of bytes transferred. There is no limit on transfer_size, but you must make sure it's safe to copy transfer_size bytes to or from the memory pointed to by buffer_pointer. A transfer_size of 1 is used to transfer a byte at a time for so-called "unbuffered" input/output.

5. lseek() : Random Access

The UNIX system file system treats an ordinary file as a sequence of bytes. No internal structure is imposed on a file by the operating system. Generally, a file is read or written sequentially i.e. from beginning to the end of the file. Sometimes sequential reading and writing is not appropriate. It may be inefficient for instance to read an entire file just to move to the end of the file to add characters. The UNIX system lets you read and write anywhere in the file. Known as random access, this capability is made possible with the lseek() system call. Random access I/O is achieved by changing the value of this file pointer using the lseek() system call.

The prototype for lseek() is :

```
long lseek(file_descriptor, offset, whence)
int file_descriptor;
long offset;
int whence;
```

Where file_descriptor identifies the I/O channel and offset and whence work together to describe how to change the file pointer according to the following table :

```
    whence              new position
------------------------------------------------------------------
    0                   offset bytes into the file
    1                   current position in the file plus offset
    2                   current end-of-file position plus offset
```

If successful, lseek() returns a long integer that defines the new file pointer value measured in bytes from the beginning of the file. If unsuccessful, the file position does not change.

For example :

```
/* lseek.c */

#include<stdio.h>
#include<fcntl.h>
int main()
  {
```

```
int fd;
long position;
fd = open("datafile.text", O_RDONLY);
if ( fd != -1)
  {
  position = lseek(fd, 0L, 2); /*seek 0 bytes from end-of-file */
  if (position != -1)
    printf("The length of datafile.text is %ld bytes.\n",
    position);
  else
    perror("lseek error");
  }
else
  printf("can't open datafile.text\n");
close(fd);
}
```

6. dup() : Channel Duplication

The dup() system call duplicates an open file descriptor and returns the new file descriptor. The new file descriptor has the following properties in common with the original file descriptor. The file descriptor refers to the same open file or pipe has the same file pointer i.e. both file descriptors share one file pointer has the same access mode, whether read, write or read and write.

The prototype for dup() is :

```
int dup(file_descriptor);
int file_descriptor;
```

Where file_descriptor is the file descriptor describing the original I/O channel returned by creat(), open(), pipe(), or dup() system calls. dup() is guaranteed to return a file descriptor with the lowest integer value available. It is because of this feature of returning the lowest unused file descriptor available that processes accomplish I/O redirection.

The following example shows standard output redirected to a file through the use of the dup() system call.

```
/* dup.c demonstrate redirection of standard output to a file. */

#include<stdio.h>
#include<fcntl.h>
#include<sys/types.h>
#include<sys/stat.h>
   int main()
{
int fd;
```

```
fd = open("anu.bar",O_WRONLY | O_CREAT, S_IREAD | S_IWRITE );
if (fd == -1)
{
perror("foo.bar");
exit (1);
   }
close(1);                    /* close standard output */
dup(fd);                     /* fd will be duplicated into standard
                             out's slot */
close(fd);                   /* close the extra slot */
printf("Hello, world!\n");  /* should go to file foo.bar */
exit (0);                    /* exit() will close the files */
   }
```

7. link() : Aliasing and Removing

The UNIX system file structure allows more than one named reference to a given file, a feature called aliasing. Making an alias to a file means that the file has more than one name, but all names of the file refer to the same data. Since all names refer to the same data, changing the contents of one file changes the contents of all aliases to that file. Aliasing a file in the UNIX system amounts to the system creating a new directory entry that contains the alias file name and then copying the i-number of a existing file to the i-number position of this new directory entry. This action is accomplished by the link() system call. The link() system call links an existing file to a new file.

The prototype for link() is :

```
int link(original_name, alias_name)
char *original_name, *alias_name;
```

Where both original_name and alias_name are character strings that name the existing and new files respectively. link() will fail and no link will be created if any of the following conditions holds :

```
A path name component is not a directory.
A path name component does not exist.
A path name component is off-limits.
Original_name does not exist.
Alias_name does exist.
Original_name is a directory and you are not the superuser.
A link is attempted across file systems.
The destination directory for alias_name is not writable.
The destination directory is on a mounted read-only file system.
```

For example :

```
/* link.c */
#include<stdio.h>
   int main()
      {
         if ((link("anu.old", "anu.new")) == -1)
            {
            perror(" ");
            exit (1); /* return a non-zero exit code on error */
            }
         exit(0);
      }
```

8. unlink() : Files

The opposite of the link() system call is the unlink() system call. unlink() removes a file by zeroing the i-number part of the file's directory entry, reducing the link count field in the file's inode by 1, and releasing the data blocks and the inode if the link count field becomes zero. unlink() is the only system call for removing a file in the UNIX system.

The prototype for unlink() is :

```
int unlink(file_name)
char *file_name;
```

Where file_name names the file to be unlinked. unlink() fails if any of the following conditions holds :

```
A path name component is not a directory.
A path name component does not exist.
A path name component is off-limits.
File_name does not exist.
File_name is a directory and you are not the superuser.
The directory for the file named by file_name is not writable.
The directory is contained in a file system mounted read-only.
```

It is important to understand that a file's contents and its inode are not discarded until all processes close the unlinked file.

For example :

```
/* unlink.c */
#include<stdio.h>
   int main()
   {
if ((unlink("anu.bar")) == -1)
   {
perror(" ");
```

```
exit (1);        /* return a non-zero exit code on error */ }
exit (0);
   }
```

9. exec : execute program

The UNIX system provides several system calls to create and end program. The UNIX system also provides several calls to send and receive software interrupts, to allocate memory, and to do other useful jobs for a process. Four system calls are provided for creating a process, ending a process and waiting for a process to complete. These system calls are fork(), the exec family, wait(), and exit(). The UNIX system calls that transform an executable binary file into a process are the exec family of system calls.

Following is a brief description of the six exec routines that make up the collective family of exec routines :

(i) execl : Takes the path name of an executable program as its first argument. The rest of the arguments are a list of command line arguments to the new program (argv[]). The list is terminated with a null pointer :

(ii) `execl("/bin/cat", "cat", "f1", "f2", (char *) 0);`

```
execl("a.out", "a.out", (char *) 0);
```

(iii) execle : Same as execl(), except that the end of the argument list is followed by a pointer to a null-terminated list of character pointers that is passed a the environment of the new program :

```
static char *env[] = {"TERM=vt100","PATH=/bin:/usr/bin",(char*)0};
execle("/bin/cat", "cat", "f1", "f2", (char *) 0, env);
```

(iv) execv : Takes the path name of an executable program (binary file) as it first argument. The second argument is a pointer to a list of character pointers (like argv[]) that is passed as command line arguments to the new program :

```
static char *args[] = { "cat", "f1", "f2", (char *) 0 };
execv("/bin/cat", args);
```

(v) execve : Same as execv(), except that a third argument is given as a pointer to a list of character pointers (like argv[]) that is passed as the environment of the new program :

```
static char *env[] = {"TERM=vt100","PATH=/bin:/usr/bin",(char*)0};
static char *args[] = {"cat","f1","f2",(char *) 0 };
execve("/bin/cat", args, env);
```

(vi) execlp : Same as execl(), except that the program name doesn't have to be a full path name, and it can be a shell program instead of an executable module :

```
execlp("ls", "ls", "-l", "/usr", (char *) 0);
```

`execlp()` searches the `PATH` environment variable to find the specified program.

(vii) execvp : Same as execv(), except that the program name doesn't have to be a full path name, and it can be a shell program instead of an executable module :

```
static char *args[] = {"cat","f1","f2",(char *) 0 };
execvp("cat", args);
```

Among the items saved by the exec system call are :

```
The "nice" value for scheduling.
The process ID and the parent process ID.
The time left until an alarm clock signal.
The current working directory and the root directory.
The file creation mask as established with umask().
All open files.
```

The last of these is the most interesting because the shell uses this feature to handle input/output redirection.

10. fork() : create a duplicate process

System call fork() is used to create processes. It takes no arguments and returns a process ID. The purpose of fork() is to create a new process, which becomes the child process of the caller. After a new child process is created, both processes will execute the next instruction following the fork() system call. Therefore, we have to distinguish the parent from the child. This can be done by testing the returned value of fork() :

- If fork() returns a negative value, the creation of a child process was unsuccessful.
- fork() returns a zero to the newly created child process.
- fork() returns a positive value, the process ID of the child process, to the parent. The returned process ID is of type pid_t defined in sys/types.h. Normally, the process ID is an integer. Moreover, a process can use function getpid() to retrieve the process ID assigned to this process.

Therefore, after the system call to fork(), a simple test can tell which process is the child. UNIX will make an exact copy of the parent's address space and give it to the child. Therefore, the parent and child processes have separate address spaces.

11. wait() : wait some process

The system call wait() is easy. This function blocks the calling process until one of its child processes exits or a signal is received. For our purpose, we shall ignore signals. wait() takes the address of an integer variable and returns the process ID of the completed process. Some flags that indicate the completion status of the child process are passed back with the integer pointer. One of the main purposes of wait() is to wait for completion of child processes.

The execution of wait() could have two possible situations.

1. If there are at least one child processes running when the call to wait() is made, the caller will be blocked until one of its child processes exits. At that moment, the caller resumes its execution.
2. If there is no child process running when the call to wait() is made, then this wait() has no effect at all. That is, it is as if no wait() is there.

12. kill() : kill a process or process id

The UNIX system sends a signal to a process when something happens, such as typing the interrupt key on a terminal, or attempting to execute an illegal instruction. Signals are also sent to a process with the kill() system call.

Its prototype is :

```
int kill (process_id, signal_name )
int process_it, signal_name;
```

Where process_id is the ID of the process to be signaled and signal_name is the signal to be sent to that process. If process_id has a positive value, that value is assumed to be the process ID of the process to whom signal_name signal is to be sent. If process_id has the value 0, then signal_name signal is sent to all processes in the sending process' process group, that is all processes that have been started from the same terminal. If process_id has the value -1 and the process executing the kill() system call is the superuser, then signal_name is sent to all processes excluding process 0 and process 1 that have the same user ID as the process executing the kill().

13. fcntl

This system call sets or clears the O_NDELAY flag, whose effect is described under write and read above.

14. fstat

The size returned is the number of bytes in the pipe, but this fact is rarely useful. A pipe may be distinguished by a link count of 0, since a pipe is the only source of a file descriptor associated with something not linked into a directory. This distinction might be useful to I/O routines that want to treat pipes specially.

15. access () : access a file

To determine if a file is accessible to a program, the access() system call may be used. Unlike any other system call that deals with permissions, access() checks the real user-ID or group-ID, not the effective ones.

The prototype for the access() system call is :

```
int access(file_name, access_mode)
char *file_name;
int access_mode;
```

Where file_name is the name of the file to which access permissions given in access_mode are to be applied. Access modes are often defined as manifest constants in /usr/include/sys/file.h.

The available modes are :

```
Value    Meaning       file.h constant
-----    ------        ------
00       existence     F_OK
01       execute       X_OK
02       write         W_OK
04       read          R_OK
```

These values may be ORed together to check for mone than one access permission. The call to access() returns 0 if the program has the given access permissions, otherwise -1 is returned and errno is set to the reason for failure. This call is somewhat useful in that it makes checking for a specific permission easy.

The following example program demonstrates the use of the access() system call to remove a file. Before removing the file, a check is made to make sure that the file exits and that it is writable.

```
/* remove.c */
#include<stdio.h>
#include<sys/file.h>
#define ERR (-1)
int main();
int main(argc, argv)
int argc;
char *argv[];
{
    if (argc != 2)
        {
        printf("Usage: %s filename\n", argv[0]);
        exit (1);
        }
    if (access (argv[1], F_OK) == ERR) /* check that file exists */
        {
        perror(argv[1]);
        exit (1);
        }
    if (access (argv[1], W_OK) == ERR) /* check for write
    permission */
        {
        fprintf(stderr,"File: %s is write protected!\n", argv[1]);
        exit (1);
        }
    if (unlink (argv[1]) == ERR)
        {
        perror(argv[1]);
        exit (1);
        }
    exit (0);
}
```

8.6 SYSTEM DEVELOPMENT TOOLS

Because UNIX System evolved in the environment of computer software development, many of the tools that form part of the system go towards assisting the program development process.

8.6.1 lint — C program checker (C- Program verifier)

Syntax

```
lint [options] filename
```

Lint detects features of C program files that are likely to be bugs, non-portable, or wasteful. It also checks type usage more strictly than the compiler. lint issues error and warning messages.

Lint produces its first output on a per-source-file basis. Complaints regarding included files are collected and printed after all source files have been processed, if -s is not specified. Finally, if the -c option is not used; information gathered from all input files is collected and checked for consistency. At this point, if it is not clear whether a complaint stems from a given source file or from one of its included files, the source filename will be printed followed by a question mark.

The behavior of the -c option allows for incremental use of lint on a set of C source files. Generally, one invokes lint once for each source file with the -c option. Each of these invocations produces a .ln file that corresponds to the .c file, and prints all messages that are about just that source file. After all the source files have been separately run through lint, it is invoked once more (without the -c option), listing all the .ln files with the needed -lx options. This will print all the inter-file inconsistencies. This scheme works well with make; it allows make to be used to lint only the source files that have been modified since the last time the set of source files were linted.

It Checks:

(a) Unreachable statement.

(b) Unused variable and argument.

(c) Unused function returns values.

(d) Ambiguous precedence operation

(e) Constructions which can make the code unportable.

The following options are used to suppress certain kinds of complaints:

-a Complaints about assignments of long values to variables that are not long.

-b Complaints about break statements that cannot be reached.

-h Do not apply heuristic tests that attempt to improve style, and reduce waste.

-m Complaints about external symbols that could be declared static.

-u Complaints about functions and external variables used and not defined, ornot used.

-v Complaints about unused arguments in functions.

-lx Include the lint library llib-lx.ln.

-n Do not check compatibility against the standard C lint library.

-p Attempt to check portability to other machines.

-s Produce one-line diagnostics only.

-c Cause lint to produce a .ln file for every .c file on the command line.

-V Write to standard error the product name and release.

8.6.2 make

make — maintain, update, and regenerate groups of programs

Syntax:

```
make [option] [File Name]
```

make allows the programmer to maintain, update, and regenerate groups of computer programs. Make is a utility oriented towards easy maintenance of computer programs. Make constructs a program or some programs according to a set of criteria held in a desription file, which is uasauly referred to as a 'makefile'. make executes commands in makefile to update one or more target names. If the -f option is not present, then makefile, Makefile, and the Source Code Control System (SCCS) files s.makefile, and s.makefiles are tried in order.

make updates a target only if its dependents are newer than the target. All prerequisite files of a target are added recursively to the list of targets. Missing files are deemed to be outdated.

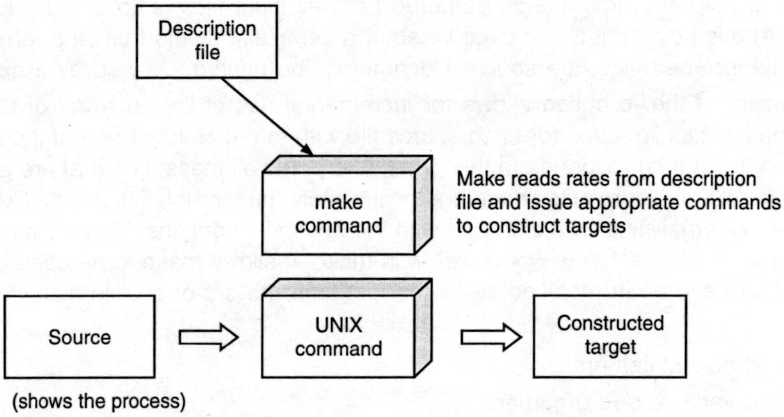

FIGURE 8.5 Flow of control in make

The makefile conations three items of information that make looks for. Three items can be considered – what, how and why

(a) The what is the name of what you are trying to construct.

(b) The how is the system commands that make should call on to construct the target from source

(c) The why is the tricky part of using make.

The options for make are listed below:

- -B Arrange the output of parallel make in appropriate blocks for readability.
- -f Description filename.
- -i Ignore error codes returned by invoked commands.
- -n No execute mode. Print commands, but do not execute them.
- -p Print out the complete set of macro definitions and target descriptions.
- -P Update in parallel more than one target at a time.
- -q Question. make returns a zero or non-zero status code depending on whether or not the target file has been updated.
- -r Do not use the built-in rules.
- -s Silent mode. Do not print command lines before executing.
- -t Touch the target files
- -w Suppress warning messages. Fatal messages will not be affected.

The following makefile says that devanu depends on five files a.o , b.o, c.o, d.o, and e.c that they in turn depend on their corresponding source files a.c, b.c, c.c, d.c, and e.c :

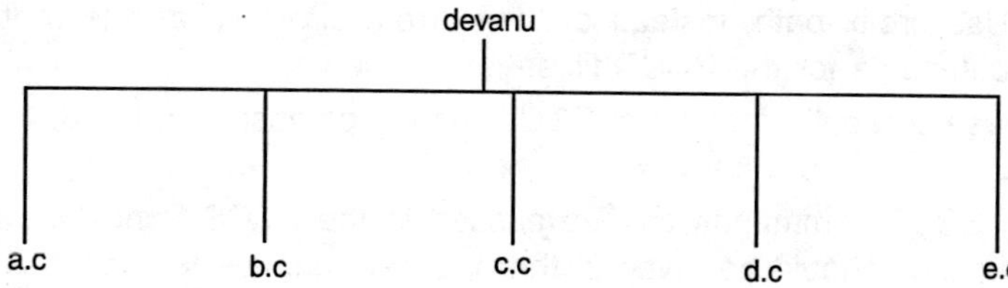


```
$ cat makefile
devanu: a.o b.o c.o d.o e.o
cc -o a.o b.o d.o e.o
a.o: a.c
cc -c a.c
b.o: b.c
cc -c b.c
c.o : c.c
cc –c : c.c
d.o : d.c
cc –c : d.c
e.o : e.c
cc –c : e.c
```

Command lines are executed one at a time, each by its own shell. The SHELL environment variable can be used to specify which shell make should use to execute commands. The default is /usr/bin/sh. The -t (touch) option updates the modified date of a file without executing any commands. Commands returning non-zero status normally terminate make. If the -i option is presented in the makefile, or if the initial character sequence of the command contains -, the error is ignored.

8.6.3 SCCS (Source Code Control System)

Syntax

```
$ sccs [command [options] [file Name ]
```

The SCCS command is a front end to various commands in the Source Code Control System (SCCS) subsystem, a tool to track changes in a file.

The database files, or SCCS files, are kept by default in a subdirectory named "SCCS" in the current directory. sccs prefixes the string "SCCS/s." to its file argument before it passes the file argument to its command argument.

SCCS does two major functions:

1. Maintains history or version control for modules.
2. Provides locks so that only one person can work on a file at one time.

Options

Sccs supports three options, -r, -d, and -p.

-r : Run command as the real user, not as the effective user that sccs has been set to through setuid. Those commands that are always run as the real user include admin, check, clean, diffs, info, rmdel, and tell.

-d prefixpath : Use prefix path, instead of the current directory, as the root directory of the pathname for the SCCS files.

-p sccspath : Use sccspath, instead of SCCS, as the pathname of the directory in which the SCCS files are kept.

The following SCCS commands can be passed to the SCCS front end, sccs. Options for these SCCS commands should be given in the options argument to sccs, following command.

admin: Create new SCCS files or change parameters of existing SCCS files. For creating new SCCS files, it is perhaps easier to use **sccs create**. Use **sccs admin** for other administrative tasks on existing SCCS files.

FIGURE 8.6 Function of admin command

For creating initial version of a file first. We place the SCCS database files in a subdirectory. We call the subdirectory devanu. We use the admin command to create the initial version of our SCCS file, which is called an s-file.

```
$mkdir devanu
$chmod +775 devanu
$admin -iarya.c devanu
```

Where arya.c is a SCCS file, which is in devanu subdirectory.

delta : Make a delta (change) to an SCCS file retrieved for editing.

get : Retrieve a version of an SCCS file. When the option -e is used, the ID keywords in the file are not expanded. The retrieved file may be edited and the changes made to the file saved later by delta. The sccs pseudo-commands sccs delget or sccs deledit combine the delta and get commands into one.

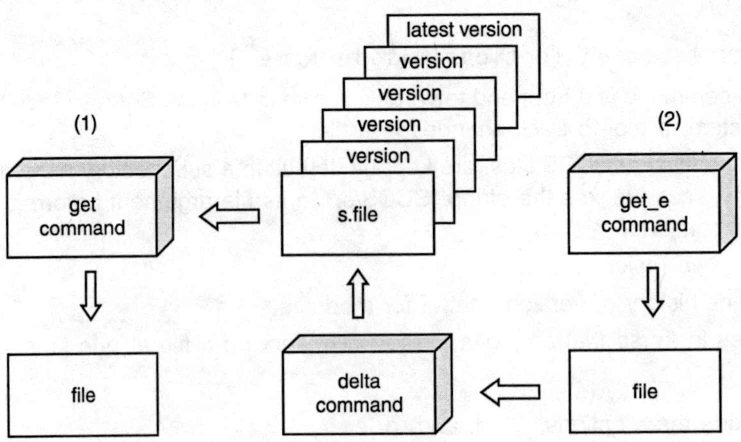

FIGURE 8.7

Path 1 uses the get command to retrieve a read read only version of the file. This is the version that you compile, or perform any other operation that does not change the file. Path 2 uses the set command with –e (for editing) option to retrieve a writable version of the file. Now you can edit the file. The general process in this path is to make your changes and check them. When you are satisfied that the change your made are correct, you then use the delta command to create a new version of the files s-file. The delta command removes the g-file.

```
$get devanu /s.arya.c
devanu /s.arya.c:
1.1
15 Lines
$get -e devanu /s.arya.c
new delta 1.2
15 lines
```

prs : Print information about an SCCS file.

rmdel : Remove a delta from an SCCS file. The delta to be removed must not be the SID of the next delta if the file is currently checked out for editing.

sact : Print a summary of editing activity for named files. The information includes the SID of the delta that currently exists in the SCCS file and the SID of the next delta, the date and time when the command **get -e** was executed, and the user who executed the command.

unget : Discard changes made to a file that is checked out for editing.

EXAMPLES

1. This example shows how to put a file named prog.c under SCCS control for the first time. First create an SCCS subdirectory if it does not exist, and then enter the file into the SCCS database, as in:

   $mkdir SCCS

   sccs create prog.c

   After verifying that prog.c has been entered successfully, remove the back up copy of the file whose name now starts with a comma:

   $rm, prog.c

2. This example shows how to get a copy of prog.c for editing, edit the file using the vi editor, and then check it back into the SCCS database, getting a read-only version of the file for later use:

   $sccs edit prog.c

   vi prog.c

   sccs delget prog.c

3. This example shows how to update a large number of SCCS database files which all have the same change in this delta, and get a read-only copy for each of the files:

   $sccs delget *.c < reason

   The text file, reason, describes the change that has been made, and provides delta with the comment text directly. The user is not asked interactively for the delta comments at the comments? prompt in this example.

8.7 TEXT FORMATTING TOOLS

- nroff
- troff
- tbl
- eqn
- pic

8.7.1 nroff (New run off) — format text

Syntax:

```
$nroff [options] FileName
```

nroff is a text formatting program that interprets source text contained in file and prepares it for printing on typewriter-like devices and line printers. The nroff command reads one or more files for printing on typewriter-like devices and line printers. If no file is specified or the - (minus sign) flag is specified as the last parameter, standard input is read by default. The File variable specifies files to be printed on a typewriter-like device by the nroff command. The default is standard input.

If the file contains plain text with no formatter requests, nroff uses default line lengths and page dimensions to produce readable output, outputting a blank line for each blank line encountered in the input, and filling and adjusting text to both margins. nroff ignores any lines in the source text that begin with a period (.) but are not valid nroff formatter requests.

Options

nroff recognizes the following command-line options, which can appear in any order but must appear before the file argument:

-olist : Print only pages whose page numbers appear in the list of numbers and ranges, separated by commas.

-nn : Number first generated page n.

-sn : Stop every n pages. nroff halts after every n pages (default n=1).

-i : Read standard input after files are exhausted.

-q : Invoke the simultaneous input-output mode of the .rd request.

-z : Print only messages generated by .tm (terminal message) requests.

-e : Produce equally-spaced words in adjusted lines.

-h : Use output tabs during horizontal spacing to speed output and reduce output character count.

-un : Set the boldening factor for the third font position (bold) to n,

-P : It allows the use of the special feature

> ✍ **Note:** These all options also use in the troff command)

Examples

The following command prints the first five pages of the document whose nroff source file is filename:

```
$nroff -o-5 filename
```

Note that there should not be a space between the o and the - or the - and the 5.

To print only pages 1, 3, and 4 type:

```
$nroff -o1,3,4 filename
```

8.7.2 troff command

Formats text for printing on typesetting devices.

Syntax:

```
$troff [options][ Filename]
```

The troff command reads one or more files and formats the text for printing on a phototypesetter or comparable device. A postprocessor is then required to post process the output of the troff command to the target device.

If no file is specified or the - (minus) flag is not the last parameter, standard input is read by default.

This process shows troff formatting a document that contains no macro- the –mm macro package is not used.

We format our document using a troff command line that like this:

```
$troff –t mscript | postprocessing program | printer command
```

troff formats the document in the file mscript, ready for driving the typesetter. We must use the –t option to direct the output to the standard output. Troff tries to drive the typesetter device.

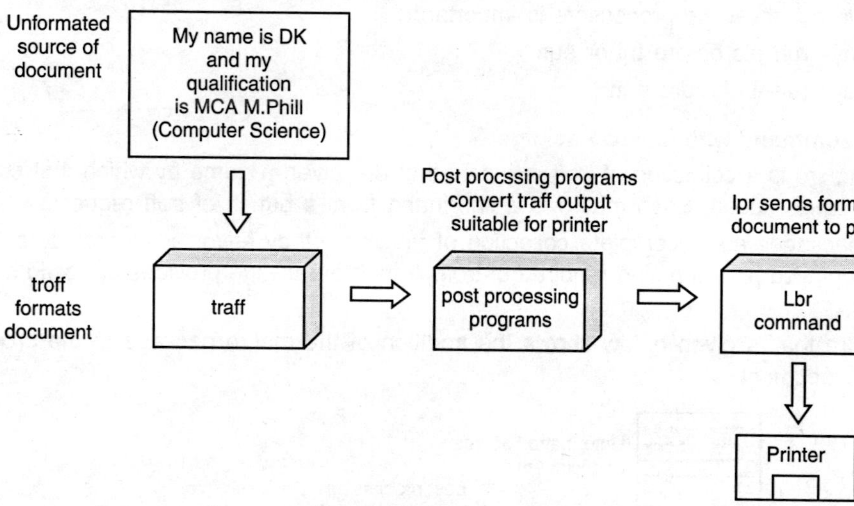

FIGURE 8.8 Basic use of troff

Using troff command with Preprocessors

The main three preprocessors available for troff are:

pic A program that interprets a language for describing pictures.

tbl A program that interprets a language for describing tables

eqn A program that interprets a language for describing equations.

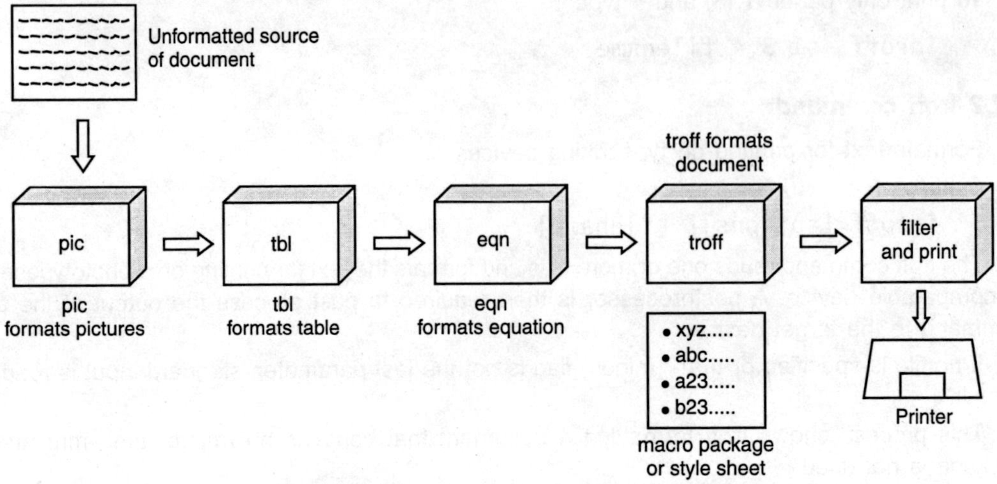

FIGURE 8.9 Using troff with preprocessors

We would use a command line like this:

```
$pic mscript | tbl | eqn | troff -mm -t | postprocessing program
\ | printer command
```

When we run these preprocessors is important :

- Always run pic before tbl or eqn.
- Always run tbl before eqn.

Using troff command with Macro Package

A troff macro is a collection of troff requests that are given a name by which that collection of requests can be used. Each macro is a shorthand from a bunch of troff requests.

Macro packages are a complete collection of macros – they envolved to provide a layer of abstraction detailed programming required of troff. A macro package provides our style sheet for a document.

The figure that is given below shows the addition of the macro package to the process of preparing a document.

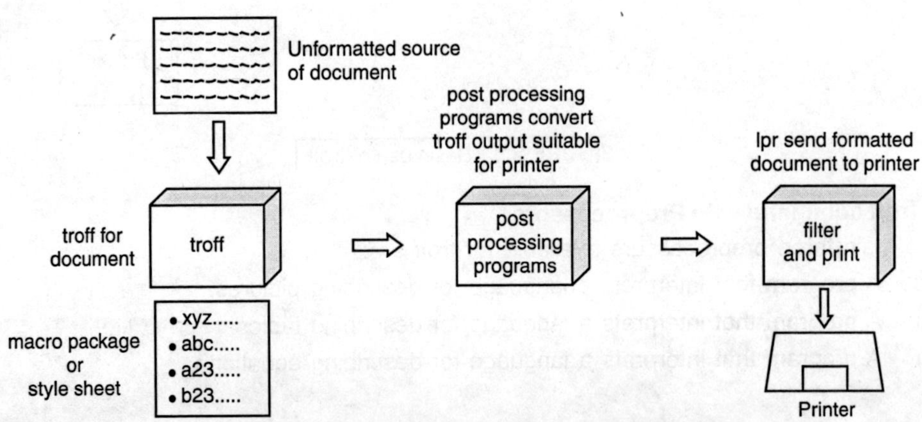

FIGURE 8.10 troff in conjunction with the mm macro package

The –mm macro package is the principle means of interfacing package, we would format our document using a troff command line that looks like this:

```
$troff -mm -t mscript | postprocessing program | printer command
```

> **Note:** We can also format these all through nroff command)

8.7.3 tbl Command

Formats tables for the nroff and troff commands.

Syntax

```
$tbl [option] [filename]
```

The tbl command is a preprocessor that formats tables for the nroff and troff commands. It reads one or more files. If no File parameter is specified as the last parameter, the command reads standard input by default. It copies the input unchanged to standard output, except for text between lines containing .TS and .TE. The tbl command reformats such text, which describes tables, without altering the .TS and .TE lines.

The tbl command processes text that is displayed within the following format:

```
.TS
          Options ;
              Format .
                  Data
.TE
```

Where TS Table Start and TE Table End

Options

Following are the available global options for the input format:

Option	Purpose
Center or CENTER	Centers the line.
Expand or EXPAND	Expands to line length.
box or BOX	Encloses in a box.
allbox or ALLBOX	Boxes all entries.
doublebox or DOUBLEBOX	Encloses in two boxes.

Format

The Format variable in the Input Format describes the format of text. Each format line describes all remaining lines of the table. A single-key letter describes each column of each line of the table.

The following are the available key letters:

Option	Meaning
l or L	Left-adjusts column.
r or R	Right-adjusts column.
c or C	Centers column.

n or N	Numerically aligns column. Numerically aligned data, n or N format
a or A	Left –adjusts subcolumn.
s or S	Spans item horizontally.
v or V	Adjusts vertical line spacing.
b or B	Indicates bold face item.
i or I	Indicates italic item.

Data

Handling data within the input format, especially for tables, uses the following line commands:

Option	Purpose
T{...T}	Indicates text block, as follows: Data<TAB>T{ Text Block T}<TAB>Data
.TS H	Used when the table continue onto more than one page.
.TH	End the header portion of the table.
.T&	Start new format.

8.7.4 eqn Command

Formats mathematical text for the troff command.

Syntax:

```
$ eqn [option][Filename]
```

The eqn command is a troff preprocessor for typesetting mathematical text on a phototypesetter or comparable device. The output of the eqn command is generally piped into the troff command.

The eqn command reads files specified by the *File* parameter. It reads standard input when a - (minus sign) is specified as the last parameter. A line beginning with the .EQ macro marks the start of equation text. The end of equation text is marked by a line beginning with the .EN macro. These lines are not altered by the troff command, so they can be defined in macro packages to provide additional formatting function such as centering and numbering.

Option

- -d : Set the ASCII characters
- -f : Change the font in all the eqn command.
- -p : Reduce subscripts and subscripts the specified no. of point in size.
- -s : Change the size in all the eqn command.
- -T : Prepares the ouput for the specified printing device.

There are two main styles of equations:

- Displayed Equations – are broken out of the main body of the text like a display.
- Inline Equations – appear right there in running line.

To use eqn with troff, we must type a command line like this:

$eqn equations | troff options | postprocessing program..........

If we want to run the other troff preprocessors, they should be run in the order:

- pic Comes first in the pipeline,
- tbl comes after pic,
- eqn comes after tble.

For example,

- EQ (Start Displayed Equation)

A sup + B sun 2 = 1

- EN (End Displayed Equation)

After formatting

$A^2 + B^2 = 1$

8.7.5 pic Command

Preprocesses troff command input for the purpose of drawing pictures.

Syntax:

```
$pic [ option ] [File name ]
```

The pic command is a troff command preprocessor for drawing simple figures on a typesetter. The basic objects are a box, circle, ellipse, line, spline, arrow, arc, and the text specified by the Text variable. The top-level object is the picture.

The top-level object in the pic command is the picture.

- PS

ElementList

- PE

If the **.PF** macro is used instead of the **.PE** macro, the position after printing is restored to what it was upon entry.

Variable names begin with a lowercase letter, followed by zero or more letters or numbers. Place names begin with an uppercase letter, followed by zero or more letters or numbers. Place and variable names retain their values from one picture to the next.

Example of pic:

1. Example, Draw ellipse and circle and connect both through a line

```
.TS
    ellipse "ellipse"
    line
    circle "circle"
.TE
```

After formatting result is

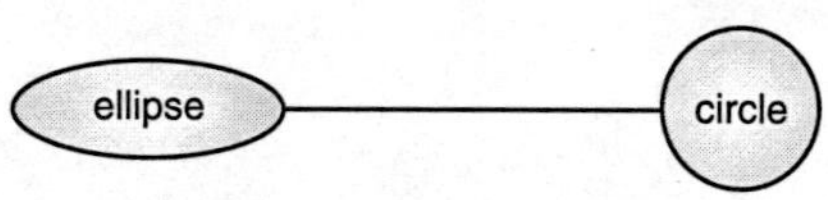

2. Example, Draw to boxes and connect them by arrow

```
.TS
box "box 1"
```

arrow
box "box 2"
.TE

After formatting result is:

☞ SUMMARY

If your application is developing new software, UNIX system is rich in tools to aid this process. In addition, the system is built around the philosophy of sharing information, re-using existing kits of parts, getting this running quickly and trying then out, and rebuilding them if they don't work out. This is possible when we use language development tools and system development tools unit make and SCCS.

In this chapter we also discuss different types of tools debugger tools for debugging such as sdb, adb, strip, ctrace, etc.

☞ EXERCISE

1. Explain the utility of Lex and Yacc in the construction of compiler. (UPTU 2001, 03)
2. What is Debugger? Name the debugger available on UNIX. What is the command to display the trace of stack with adb debugger?
3.What are different language development tools? Btiefly explain each of them.
4. What do you understand by debugger tools? How does trace help in debugger? Explain with examples.
5. What are the difference among a system call, a library function and a UNIX command?
6. What is system call? Discuss any four system calls.
7. Describe the system development tools :
   (i) LINT (ii) SCCS (iii) make
8. How does make work?
9. What do you mean by System Development Tools? Name any two tools and explain then in short.

❑❑❑

ADDITIONAL PROGRAMS (SCRIPTS)

Program 1: Write a shell program to reverse the digits of an input number n.

```
echo enter the number for reverse
read n
while test $n - gt 0
do
    echo ‘expr $n % 10’
    n = ‘expr  $n / 10’
done
```

Program 2: Write a shell program that accepts one or more files names os arguments and converts their contents to upper case.

```
for i in $*
do
    cat $i ¦ tr “[a-z]” “[A-Z]” > T
    cat T > $i
    echo the convert file is $i
    cat $i
done
```

Program 3: Write a shell program to generate a table of an input number.

```
echo “Enter the number for table”
read n
a = 1
echo “The generate table is =”
while test $a - le 10
do
    b = ‘expr $a \* $n’
    echo $b
    a = ‘expr $a + 1’
done
```

Program 4: Write a shell program to find out the a power b.

```
echo "Enter the value of a & b
read a b
c = 1
d = 1
while test $c – le $b
do
    d = 'expr $d \* a'
    c = 'expr $c + 1'
done
echo "The power "$a" of "$b" is "$d"
```

Program 5: Write a shell program to find out the factorial

```
echo "Enter the number for factorial
read n
i = $n
a = 1
while test $n - ge 1
do
    a = 'expr $a \* $n'
    n = 'expr $n - 1'
done
echo "The factorial of "$i" is "$a".
```

Program 6: Write a shell program to find out the even or odd.

```
Echo "Enter the no. of check no even or odd"
read n
if test 'expr %2' – eq 0
then
    echo "No. is even"
else
    Echo "No is odd"
fi
```

Program 7: Write a shell program to find out greatest among three numbers.

```
Echo "Enter any three number"
read a b c
    if test $a -gt b
then
    if test $a -gt c
then
```

```
    echo $a "is greatest numbner"
else
    echo $c "is greatest number"
fi
    elseif test $b -gt $c
then
    ehco $b "is greatest number"
else
    echo $c "is greatest number"
fi
```

Program 8: Write a shell program to find the length of string

```
Echo "Enter the string
read s
    l = 'echo $s ¦ wc -c'
    l = 'expr $l - 1'
echo "Length of the string is $l
```

Program 9: Write a shell program to compare two strings.

```
Echo "Enter two strings for compare
read s1s2
if test $s1-eq$s2 $s1 – eq $s2
then
echo "string are equal"
else
echo "string are not equal"
```

Program 10: Write a shell program to check whether a string is palindrome or not.

```
Echo "Enter any string for check it is palindrome or not"
read s
l = 'echo $s ¦ wc–c'
l = 'expr $l – 1'
f = 0
a = 1
while test $a -le $l
do
    b = 'echo $s ¦ cut -c$a'
    c = 'echo $s ¦ cut -c$l'
    a = 'expr $a+1'
    l = 'expr $l–1'
if test $b - eq $c
then
```

```
f = 0
else
f = 1
break
fi
done
    if test $f – eq 1
echo "string is not Palindrome
else
    echo "string is Palindrome
fi
```

Program 11: Write a shell program to concastinate two files.

```
Echo "Enter two files
read arya anu
cat arya anu>dk
cat dk
```

Program 12: Write a shell program to find out the sum of series 1 + 2 + 3 + ... n, where n is input by the user.

```
add=0
i=1
echo "Enter your Limit:/c"
read n
while test $i -le $n
do
    add='axpr $add + $i
    i='expr $i + 1'
done
echo "sum of $n Natural Numbers is $add"
```

Program 13: Write a shell program to generate the series 2, 4, 6, 8, ... n. where n must be input by the user.

```
i=2
echo "Enter the limit for series :/c"
read n
while test $i -le $n
do
    echo "$i/
    i='expr $i + 2'
done
echo "/b/b"
```

Program 14: Write a shell program to generate the series 1, 5, 2, 10, 3, 15, 4, 20, 5, 25, 6, 30, ..., 10, 50.

```
i=1
j=5
p=0
while test $i -le 10
 do
    echo "$i"
    p='expr $j/*$i
    echo "$p,/c"
    j='expr $i + 1'
done
echo "/b."
```

Program 15: Write a shell program to find out the sum of series 1 + 1/2 + 1/3 + ... + 1/N.

```
i=1
s=0
whole=0
frac=0
div=1
echo "Enter the value for N :/c"
read n
while test $i -le $n
 do
    q='expr $div/$i
    s='expr $s + $q'
    s='expr $s/*10'
    div=expr $div/*10'
    i='expr $i + 1'
done
whole='expr $s/$div'
trac='expr$s%$div'
echo
echo "Sum of Series (1 + 1/2 + 1/3 + .... 1/$n) = $whole.$frac"
```

Program 16: Write a shell program to find out the sum of series $1^2 + 2^2 + 3^2 + \ldots\ldots\ldots\ldots n^2$.

```
i=1
p=0
sum=0
echo "Enter the limit :/c"
read n
```

```
while test $i -le $n
 do
     p='expr $i/*$i'
     sum='expr $sum + $p'
     i='expr $i + 1'
 done
echo "Sum of series : $sum"
```

Program 17: Write a shell program to find out the greatest among the three inputs.

```
max=0
echo "Enter the first No. :/c"
read a
echo "Enter the second No. : /c"
read b
echo "Enter the third No. : /c"
read c
if test $a -ge $b
 then
     max=$a
else
     max=$b
fi
if test $max -ge $c
 then
     echo "$max is the greatest input among all three inputs."
else
     echo "$c is the greatest input among all three inputs."
```

Program 18: Write a shell program to find out the greatest among n inputs integers where n is too inputs by the user.

```
n=0
i=1
rm list
touch list
rm temp
touch temp
echo "Enter the value for n : /c"
read n
while test $i -le $n
 do
     echo "Enter the any integer : /c"
```

```
        read no
        echo $no >> list
        i='expr $i + 1'
    done
    sort list > temp
    echo "Greatest Number amongst $n integer =/c"
    tail -1 temp
```

Program 19: Write a shell program to concatenate two strings input by the user.

```
echo "Enter Your First Name :/c"
read str1
echo "Enter Your Last Name :/c"
read str2
echo $str1 $str2 > string
echo "Concatenated String :/c"
cat string
```

Program 20: Write a shell program to read an integer and print it in reverse order.

```
no=0
mod=0
rev=0
div=10
echo "Enter any No. :/c"
read no
echo "Reverse No. of $no : /c"
while test $no -gt 0
 do
    mod='expr $no %10.
    rev='expr $rev /*10'
    rev='expr $rev + $mod
    no='expr $no /10'
done
```

Program 21: Write a shell program an arithmetic operation upon two inputs. The operation should also input by the user.

```
ans=0
ch=0
echo "Enter the first No.: \c"
read a
echo
echo "Enter the second No. : \c"
read b
```

```
echo
echo "Select any one choice :-"
echo
echo "Enter 1 for addition."
echo
echo "Enter 2 for Subtraction."
echo
echo "Enter 3 for Multiplication."
echo
echo "Enter 4 for Division."
echo
echo "Enter Your choice : \c"
read ch
echo
case $ch in
1) ans='expr $a + $b'
echo "$a + $b = $ans";;
2) ans= expr$a - $b'
echo "$a - $b = $ans";;
3) ans='expr $a\* $b"
echo "$a * $b = $ans";;
4)-if test $b -eq 0
then
    ans='expr $b / $a'
    echo "$b / $a = $ans"
else
    ans='expr $a/$b'
    echo "$a / $b = $ans"
ft;;
*) echo "Sorry! You have entered a Wrong choice."
esac
```

Program 22: Write a shell program to find out the length of an input string.

```
clear
echo "Enter any string : /c"
read str
len='expr $str/wc -c'
len='expr $len -1'
echo "Length of $str is $len"
```

Program 23: Sort the given numbers in the given order i.e., either ascending or descending order. (Use "SORT" command of unix.)

```
lim=0
i=1
ch=0
rm file1
touch file1
rm file2
touch file2
echo "Enter the No. of elements to be sort :/c"
read lim
while test $i -le $lim
 do
    echo "Enter any Number : /c"
    read list
    echo $list >> file 1
    i='expr $i + 1'
 done
echo "Select any one choice :"
echo "Enter 1 for Sorting in ascending order."
echo "Enter 2 for Sorting in deceding order."
echo "Enter Your Choice : \c"
read ch
echo "Your Choice is $ch"
if test $ch -eq1
 then
    sort file1 > file2
    echo "Ascending Order Sorting;-"
    cat file2
else
 if test $ch -eq 2
then
    sort-r file1 > file2
    echo "Descending Oder Sorting :-"
    cat file2
else
    echo "Sorry! ..."
    fi
fi
```

Program 24: Write a shell program to print an input string into its reverse order.

```
echo "Enter any string :/c"
read s1
len='expr $s1/wc -c'
len='expr $len -1'
while test $len -gt 0
 do
     r='echo $s1| cut -c $len'
     echo "$r/c"
     len 'expr $len -1'
  done
echo
```

Program 25: Write a shell program to find out the location of an input character into an input string.

```
i=1
echo "Enter a string :/c"
read s1
len='expr $s1 |wc -c'
len='expr $len -1'
echo "Enter the character U want to search :/c"
read ch
while test $i -le $len
 do
     k=echo $s1 | cut -c $i
     if test $k = $ch
     then
          echo "Position of $ch in the string is $i"
          break
fi
     i'expr $i + 1'
done
if test $i -gt $len
then
     echo "Character does not exist in the String."
fi
```

Program 26: Write a shell program to find whether an input year is leap year or not.

```
echo "Enter any Year :/c"
read year
m='expr $year %4'
 then
```

```
if test $m -eq 0
then
    echo
    echo "$year is a Leap Year."
else
    echo
    echo "$year is not a leap year."
fi
else
    echo
    echo "Wrong Year Entered by You."
fi
```

Program 27: Write a shell program to translate a file into UPPER CASE through command line.

```
if test $# -eq 0
 then
    echo "Sorry No Argument Passed By User."
else
 if test $# -gt1
  then
    echo "You have passed more than one Argument."
    echo
    echo "But Only the first argument is accepted as file name i.e. $1"
    echo
else
    echo "File Name Entered By The User is : $1"
    echo
    echo "Now Enter Any Key to Continue"
    read xx
    temp="$1temp"
    rm $temp
    touch temp
    clear
fi
    echo "Now Enter text in the file and save it."
    echo
    cat>$1
    tr "a-z" "A-Z" < $1 > temp
    cat temp > $1
    rm temp
    echo "/n"
```

```
    echo "Text of $1 File is translated into the UPPER CASE|-"
    echo
    cat $1
fi
```

Program 28: Make a duplicate copy of a specified file through command line.

```
if test $# -eq 0
 then
    echo "Sorry! No Argument Passed By User."
else
if test $# -gt 2
 then
    echo "You have passed more than Two Argument."
    echo
    echo "But Only the first two arguments are accepted."
    echo
fi
    echo "Source File Name Entered By The User is : $1"
    echo
    echo "Target File Name Entered By The User is : $2"
    echo
    cp $1 $2
    echo
fi
```

Program 29: Write a shell program to find the odd and even numbers in a given list of numbers.

```
i=1
mod=0
echo "Enter the Limit of List : /c"
read lim
while test $i-le $lim
 do
    echo "/n Enter the Value for list[$i] : /c"
    eval read list$i
    i='expr $i + 1'
done
i=1
clear
echo "List Contains the Following Elements :-"
while test $i -le $lim
 do
```

```
        echo
        eval echo "/$list$i"
        i='expr $i + 1'
  done
i=1
echo
while test $i -le $lim
  do
        eval mode='expr "/$list$i" %2'
        if test $mod -eq 0
         then
                eval echo "/$list$i" is a Even Number.
                echo
else
            eval echo "/$list$i" is a Odd Number.
            echo
fi
i='expr $i + 1'
```

□□□

MODEL QUESTION PAPERS

MODEL QUESTION PAPERS-I

MODEL QUESTION PAPER-I

UNIX AND SHELL PROGRAMMING

Time : 3 Hours　　　　　　　　　　　　　　　　　　　　Total Marks: 100

Notes : (1) Attempt all the questions

(2) All questions carry equal marks.

1. Attempt any FOUR parts of the following: **(5×4=20)**

(a) Draw the Architecture of UNIX system and describe each layer.

(b) Differentiate between the Background process and foreground process. Also give the proper example for each.

(c) Give the syntax of the following commands and explain the uses also :

(i) fsck

(ii) chown

(iii) PID

(iv) diff

(d) What do you mean by Sleep and Wakeup in UNIX?

(e) What is the purpose of permissions in UNIX file system? Explain the different levels of permissions. Also differentiate between the ordinary files and special files in UNIX operating system.

(f) Write the commands to perform the following functions :

(i) Display the directory page by page.

(ii) Find out how many users are presently logged.

(iii) Count the number of words and characters in a file.

(iv) Delete all files with names starting with "k".

(v) List all the processes running on the system.

2. Attempt any FOUR Parts of the following: **(5×4=20)**

(a) What are the modes of execution in UNIX ? Explain the services of any one mode.

(b) What do you mean by the Shell? Explain different types of shell that are available in UNIX.

(c) What do you mean by the term, Process? Explain it by the Process state transition diagram.

(d) How will you remove a directory tree even when it is not empty without using rmdir?

(e) What are the important features of UNIX operating system ?

(f) What are the role of KERNEL in UNIX operating system? Explain the main features of KERNEL.

3. Attempt any TWO Parts of the following: **(10×2=20)**

(a) How does **make** work ? Explain it.

(b) What facilities are used for TEXT formatting in UNIX operating system?

(c) What do you mean by System Development Tools? Name any two tools and explain them in short.

4. Attempt any TWO Parts of the following: (10×2=20)

(a) Write a shell script using **date** that prints the usual date output as default but which has options for printing just the time, just the day-month-year, or just the day of the week.

(b) Determine what the built-in shell variables are for your account. Which ones are exported?

(c) Find out the result of the following :

(i)
```
For i in x y z
do
echi $i
done
```

(ii)
```
integer a, b, c
a = 280  b = 257  c = 514
((a=a+b/c));
print $a
```

5. Attempt any TWO Parts of the following: (10×2=20)

(a) What do you mean by a 'Daemon'? Explain the function of **init** daemon. How will you kill a daemon?

(b) Write shore notes on the following :

(i) pipe

(ii) message queue

(iii) shared variable

(iv) system call

(c) What are the differences among a System call, a Library function and a UNIX command?

❑❑❑

MODEL QUESTION PAPER-II

UNIX AND SHELL PROGRAMMING

Time : 3 Hours Total Marks: 100

Notes : (1) Attempt all the questions

(2) All questions carry equal marks.

1. Attempt any FOUR parts of the following: (5×4=20)

(a) Sketch a diagram showing the components of UNIX and discuss kernel-shell relationship.

(b) Discuss different operational modes of vi editor. Write vi commands to perform the following :

(i) Generate line numbers for the file being edited currently.

(ii) Delete three lines downward from current line.

(iii) Mark a block of lines and perform cut and paste.

(iv) Save the currently opened file with a new name.

(v) Replace a word with new text.

(c) What is system call? Discuss any four system calls.

(d) What is the triad of file permissions? Discuss their effect on regular and directory files.

(e) "UNIX views everything in the form of file". Justify this statement with suitable examples.

(f) Why is UNIX more portable than any other operating system? Discuss its multi-user and multi-tasking features.

2. Attempt any FOUR Parts of the following: (5×4=20)

(a) Discuss the utility of shell metacharacters using suitable examples.

(b) Compare and contrast the system and user variables. Discuss any five system variables of Bourne shell.

(c) Write a shell program to reverse the digits of an input numnt "n".

(d) Explain different shell programming constructs with suitable examples.

(e) Write a shell program that accepts one or more file names as arguments and converts their contents to upper case.

(f) (i) Explain the usage of following variables :

(a) S$ (b) $* (c) $$ (d) PSI (e) $?

(ii) Discuss i-node of UNIX operating system.

3. Attempt any TWO Parts of the following: (10×2=20)

(a) (i) Differentiate between Orphan and Zombie processes. Write program(s) create Orphan and Zombie processes.

(ii) What is semaphore? Discuss its importance in multiprogramming environment.

(iii) What do you understand by log-in cycle? Discuss it with the help of suitable example.

(b) (i) Write the command/command sequence to :

(a) Find all logged-in users having names of at least four characters.

(b) Find the number of users on your system whose user-id is greater then 200.

(ii) Write a 'C' program which forks a child process and overwrites it using exec system call with code printing "UPTU, Lucknow.

(c) Why do we use Sockets ? Discuss the following system calls of socket mechanism :
 (i) socket ( ) (ii) bind ( ) (iii) connect ( ) (iv) listen ( ) (v) accept ( )

4. Attempt any TWO Parts of the following: (10×2=20)

(a) (i) What is file system? Discuss the creation, mounting and unmounting of file system in UNIX.
 (ii) Discuss the usage of "at", "batch" and "cron".

(b) (i) Explain briefly the use of the following files :
 (A) profile (B) /etc/profile (C) /etc/group
 (D) /etc/passcod (E) m/etc/default/msdos

(c) (i) Discuss using suitable examples the backup and restoration of files using cpio. How is it different from tar?
 (ii) How are the blocks of file system organized in four categories? Explain each of them.

5. Attempt any TWO Parts of the following: (10×2=20)

(a) Discuss the following system development tools in brief.
 (i) lint
 (ii) make
 (iii) SCCS

(b) What are different language development tools? Briefly explain each of them.

(c) (i) What do you understand by debugger tools? How does trace help in debugging? Explain with examples.
 (ii) How can the text formatting tools, like troff be used for text formatting? Discuss salient features of troff or nroff.

□□□

MODEL QUESTION PAPER-III

UNIX AND SHELL PROGRAMMING

Time : 3 Hours Total Marks: 100

Notes : (i) Attempt ALL the questions

(ii) All questions carry equal marks.

(iii) In case of numerical problems assume data wherever not provided.

(iv) Write your answer as concisely as possible, consistent with providing a complete answer to the question.

(v) Be precise in your answer.

1. Attempt any FOUR parts of the following: (5×4=20)

(a) What do you mean by multi-user and multitasking OS?

(b) Explain in details the Directory entry, inode entry and file contents.

(c) You create a file called myfile in the directory mydir. Later you add new material to myfile. What changes take place in directory, inode and file? What command would give you read, write and execute permissions, group members read and execute permissions and other just permissions to read the file myfile?

(d) Explain the UNIX file system.

(e) Which command is used to write in core super block and inode table information to the disk? Can a file have more then two links? Can a file have links across the file systems?

(f) What will be the effect of following UNIX commands :

(i) unmask 022 (ii) ulimit 512 (iii) du../..

(iv) du. (v) in dir 1 dir 2 (dir 2 is existing on disk)

2. Attempt any TWO Parts of the following : (10×2=20)

(a) Write a shell script to count the report the number of entries present in each sub-directory mentioned in the path which is supplied as command line argument.

(b) (i) Write a shell program to find the factorial value of any number entered through keyboard.

(ii) Two numbers are entered through keyboard. Write a shell program to find the value of one number raised to the power of another.

(c) Write a shell script which works similar to the following UNIX commands :

head tail more

3. Attempt any TWO Parts of the following: (10×2=20)

(a) (i) Write a script that takes a command-line argument (which is a name of a file in the present working directory) and reports on whether it is a directory, a file, a something else.

(ii) Write a script that will check every 30 seconds if a user called "Ramesh" has logged into system. If Ramesh is logged in when the script is first run, the script will terminate, saying "Ramesh is logged in." Otherwise, it will check every 30 seconds to see if Ramesh has logged in, and when has, script will terminate with the same message.

(b) Consider the following script called **Journal_add:** # journal : add journal entries to a $HOME/journal

```
File = $HOME/journal
date >> $file
echo "Enter name of persion or group :\c"
read person
echo "$person" >> $file
echo >> $file
cat >> $file
echo >>" ------------" >> $file
echo >> $file
```

Add commands to **journal_add** to verify the user has write permissions for a file called **journal** in the user's home directory, if such a file exists. The script should take appropriate actions if a **journal** file exists and the user does not have write permission. Why does the script use **read** the first time it accepts input from the terminal and **cat** the second time?

(c) Explain the following :
   (i) Semaphore
   (ii) Pipes
   (iii) Socket Programming

4. Attempt any FOUR Parts of the following: **(5×4=20)**

(a) Write a shell script to perform the following operations.
   (i) Take backup of files in/usr/cprogs directory at 6:00 PM every day.
   (ii) List all backedup file present on a floppy disk.

(b) Outline the jobs of a system administrator.

(c) What is the use of a tar command and how is it different from other UNIX commands?

(d) Write an interactive shell script which will request the user to put the 1.2 MB floppy in drive A, format it, make a file system on it with 2000 blocks and 500 inodes, mount it at a mount point /dev/fdd.

(e) State whether the following statements are True or False
   (i) A file can be copied across the two different file systems.
   (ii) A file system can be mounted and un-mounted by an user.
   (iii) The command /etc/init 0 halts the system.

(f) Write a shell script to send mail to groups of users by extracting their ID's from /etc/**group** file.

5. Attempt any FOUR Parts of the following: **(5×4=20)**

(a) What is the use of Makefile and **make** command?

(b) Write short notes on lint, Source code control System and YACC.

(c) Explain the usage of various debuggine tools. What is the use of **ctrace?**

(d) What is the difference between a pipe and redirection operator.

(e) Write a script called **del** which takes a file name as argument and ask you whether you want to delete the file of not.

(f) "UNIX is a multi-user multitasking Operating system". Is it true? Discuss in details.

❏❏❏

MODEL QUESTION PAPER-IV

UNIX AND SHELL PROGRAMMING

Time : 3 Hours Total Marks: 100

Notes : (i) Attempt ALL the questions

(ii) All questions carry equal marks.

1. Attempt any FOUR parts of the following: **(5×4=20)**

(a) Draw the architecture of UNIX system and describe each layer.

(b) What do you mean by sleep and wakeup in UNIX?

(c) Write the commands for the following :

(i) Display the calender of April 2007.

(ii) List all processes running on the system.

(iii) Count the number of words and characters in the file

(iv) To find the name of your machine.

(v) Checking the spelling of the phrase "Hello World".

(d) Give the complete syntax of the following commands

(a) ls

(b) date

(c) time

(d) rm

(e) mkdir

(e) Write a short note on internal and external commands of UNIX.

(f) What are the various methods to get help in UNIX ?

2. Attempt any FOUR Parts of the following: **(5×4=20)**

(a) What is the role of Kernel in UNIX operating system? Explain the various features of KERNEL.

(b) Write a short note on filing mounting and unmounting in UNIX.

(c) How do you create a new user account in UNIX considering that you are the administrator? Name this account as ABCXYZ and set its password as NULL VALUE (The user decides the password once s/he logs on for the first time).

(d) What is the various security features offered in UNIX ? Explain any two in detail.

(e) Program a UNIX machine to handle ten logins simultaneously. Name them as mac1, mac2, mac3... mac10. All these login terminals should have the same password as "Spiderman" and access rights should also be similar.

(f) Write commands to backup all the data from the account "HELLRAIZER" everyday at 22.30 hrs.

3. Attempt any TWO Parts of the following: **(10×2=20)**

(a) Write a short note on language development tools in UNIX.

(b) Describe the following tools troff, nroff, tbi.

(c) Describe SCCS with an example.

4. **Attempt any FOUR Parts of the following :** **(5×4=20)**
   (a) Distinguish between Bourne shell, C shell and Korn shell.
   (b) Write a short note on arrays in UNIX.
   (c) Determine what the built-in shell variables are for your account. Which ones are exported.
   (d) Write a shell script to print "Hello World" in inverted colours on the screen.
   (e) Explain all the decision making statements in details.
   (f) How will you repeat the last command in
      (i) Korn Shell
      (ii) Bash ?

5. **Attempt any TWO Parts of the following:** **(10×2=20)**
   (a) Write a shell script that prints the RAM used by the system.
   (b) Write a shell script that gives the memory details of the system.
   (c) Explain Multitasking, background processes and child-parent relationship in details.

❑❑❑

INDEX

INDEX

A

Accessing Array Values, 164
Adb (Absolute Debugger), 243
Adding Group: groupadd, 175
Addresses, 194
Addressing Lines, 40
Advantage and Disadvantage of
- Shell Script, 140
- UNIX File System Organisation, 27
- UNIX
Application Programs, 12
Arithmetic
- Expressions, 136
- Operators, 156
Array Values, 164
Arrays, 16,4 203, 220
awk Filtering, 195
AWK, 195

B

Back
- Array, 164Quotes, 139
- Slash, 140
Backup and Recovery, 176
Basic File Attributes, 18
Boolean Operators, 157
Boot Block (BB), 23
Boot Phases, 172
Bracket ([ ]) Wildcard Character, 128
Break Statement, 155

C

case- esac Statement, 150
Changing Your
- Login Shell, 134
- Password, 14
Chmod, 181
Command Grouping: (Parentheses), 139
Command Line Interpretation, 166
Comments, 194
Comparison Operators, 199
Conditional Execution Using && and ||, 138
Continue Statement, 155
Control Flow, 206
Controlling Terminal Inputs, 159
Copying Text, 35
CPIO: Copy Input Output, 176
Creating Files, 182
Ctrace - C program debugger, 244

D

Data
- Block, 24
- Data Types, 215
Dbx, 242
Debugger Tools, 242
Decision making, 141
Deleting Group : groupdel, 175
Deleting Text, 33
df : For find the free Disk Space184
dfspace : Fee Disk Space int MB, 185
Directory Files, 17
Disk Related Commands, 184
Double Quotes, 139
du : Used Disk Space, 185
dump Command, 179

E

ed Editor, 37
Editing a File, 43
Editing With ed, 73
Encrypting Files, 182
Entering Text, 33
Entering Your Userid and Password, 13
Environment, 134
eqn Command, 268
Examples of Interrupt Handling, 162
Execute file, script or program, 37

F

Features of ed, 75
Features of UNIX Shell, 126

File System Mounting and Unmounting, 170
File System, 11, 19
File Test Operators, 158
Files, 16
Finding Information about Library Functions, 245
Flags, 38
for loop, 152
for Statement, 235
foreach Statement, 236
Forming Patterns, 39
Functions of Operating System, 3
Functions, 204, 227

G

General purpose utilities, 78
Generalized Regular Expression Parser, 190
Getting Started, 30

H

Handling Signals, 161
Handling User Account, 173
Hardware, 12
History of UNIX, 6

I

if-then- fi statement, 141
if-then-elif-else-fi statement, 146
if-then-else-fi-statement, 142
Inode (Index Nodes), 24
Input for GREP, 190
Input Redirection from a File, 160
Inserting Comments, 139
Integer Arithmetic, 135
Internal and External Commands, 78

K

Kernel, 10

L

Language Development Tools, 238
Lexical Analyzer Generator, 239
Library Functions, 245
lint–C program checker (C-Program verifier), 259
Logging In and Out, 13
Logging Out, 14
Looping Control, 152

M

Main Features of UNIX, 12
make, 259
Manipulating Files, 68
Miscellaneous Functions of the ed Editor Subcommands, 70
Modifying the Group: groupmod, 176
mount : Mount File System, 171
Moving Text, 35
Moving the Cursor, 32

N

nroff (New run off) — format text, 264

O

Operating System, 1
Operators, 222
Options in GREP, 191
Output Redirection to a File, 159
Overview of UNIX, 4

P

Passwd: Changing Password, 181
Pattern Matching, 38
PERL, 212
pic Command, 269
Positional Parameters: Set, 133
printf: Formatting Output, 197
Producing the Output, 140

Q

Quoting Metacharacters, 129

R

Read Statement, 132
Regular Expressions, 39
Regular Files or Ordinary Files, 17
Relational Operators, 157
Replace and Substitute, 34
Reserved Words, 167

Restricting: Pattern matching, 40
Root Directory (/), 18

S

SCCS (Source Code Control System), 261
Scrolling Text, 36
Sdb (Symbolic Debugger), 243
Searching a Pattern, 190
Searching for Patterns in vi, 37
Security, 180
Setting User ID (SUID) Permission, 180
Shell, 11
- Commands, 133
- Meta-Character (Wildcards), 127
- Operators, 156
- Script, 132
- Variables, 129

Shifting Arguments: Shift, 133
Shutting down, 173
Simple Shell Script, 133
Single Quotes, 139
Some
- Common System Administration Task, 169
- Standard Shell Variables, 131
- Useful Library Functions, 246
- Useful System Calls, 246

Space Balls, 195
Special
- Command Line Characters, 138
- Files or Device Files, 17
- Shell Variables, 131

Splitting a Line into Fields, 196
Sticky Bit, 181
Storage of Files, 183
Stream Editor (SED), 192
String
- Operators, 158
- Manipulating, 137

strip, 243
Subcommands, 43
Substitution Command, 194
Super Block (SB), 24
System
- Booting, 172
- Calls, 246
- Development Tools, 258
- State Summary (or Init Level), 172

T

tar Command: the tape Archive Program, 178
tbl Command, 267
Text Formatting Tools, 264
The * Wildcard, 127
The ? Wildcard, 128
trap: Handling command lists, 163
Trapping signals, 160
troff command, 265
Types of
- File Systems, 20
- Operating System, 2
- Shells, 125

U

ulimit : User Limit, 186
umount: Unmount File System, 171
Unix Directories, 22
Unix File Structure, 17
Unix File System Organization, 23
Unix Signals, 160
Unix System Organization, 10
Until Loop, 154
Use of Library Functions, 245
User Accounts, 187
User Quatas, 186
useradd: Adding new user, 174
userdel & usermod: Removing and modifying users, 12, 175
Utilities, 11

V

Variables and Expressions, 198
Variables, 202, 218
vi Modes, 29

W

While Loop, 153
while/until Statement, 236
Writing PERL Program, 214

Y

Yacc-Yet another Compiler Compiler a parser generator, 238

❑❑❑

Win Prizes

ATTENTION READERS

Win Prizes

We request for your frank assessment regarding various aspects of the book as given below :

UNIX AND SHELL PROGRAMMING

By: Devendra Kumar

Please Fill up the Sheet in Neat and Clean Writing

(i) What topics of your syllabus, important for your examination point of view, are not covered in this book ?

....

....

....

....

(ii) What are the chapters wherein the treatment of the subject matter is not systematic or organised or updated ?

....

....

....

....

(iii) Have you come across misprints/mistakes/factual inaccuracies in the book? Please specify the chapters and the page numbers.

....

....

....

....

(iv) What other books on the same subject have you found/heard better than the present book? Please specify in terms of price and quality

....

....

....

....

(v) Further suggestions and comments for the improvement of this book.

....

....

....

....

(PS: If need be please attach a separate sheet to write your views)

OTHER DETAILS

(i) Who recommended you this book ? (Please tick (✓) in the box near the option relevant to you.

☐ Teacher ☐ Friends ☐ Bookseller

....

....

....

....

(ii) Name of the recommending teacher, his designation and address :

....

....

....

....

(iii) Name and address of the bookseller you purchased the book from :

....

....

....

....

(iv) Name and address of your institution (Please mention the University or Board as the case may be)

....

....

....

....

(v) Your name and complete postal address :

....

....

....

....

(vi) Course you are enrolled in :

....

....

....

....

The best assessment will be awarded each month: The award will be in the form of our publications as decided by the Editorial Board.

Please mail the filled up coupon at your earliest to :

S.K.KATARIA & SONS®
4885/109, Prakash Mahal, Dr. Subhash Bhargav Lane,
Opposite Delhi Medical Association, Daryaganj, New Delhi–110002 (INDIA)
Phone: +91-11-23243489, 23269324, 43551243; *Telefax:* +91-11-23243489
e-mail: katariabooks@yahoo.com; katariabook@gmail.com
Website: www.skkatariaandsons.com; *For e-books:* www.katsonbooks.com